AF862020

Learn and Do Microscope Adventure Unit Study

For information, email info@KymWright.com or
Kym Wright
PO Box 81124
Conyers, Ga 30013.

Websites:

www.KymWright.com

www.Learn-and-Do.com

www.Living-Life-on-Purpose.com

www.The-Mothers-Heart.com

International Standard Book Number: 978-1-4276-2074-3

Contents

People to Study

Marcello Malpighi
Galileo Galilei
Hans Lippershey
Hans Janssen
Zacharias Janssen
Francesco Stelluti
Robert Hooke
Matthias Schleiden
Theodor Schwann
Antony van Leeuwenhoek
J. J. Lister
Louis Pasteur
Robert Koch
Ernst Ruska
Max Knoll
Frits Zernike
Rudolph Virchow

How to Use *Microscope Adventure!*

Lab Materials

Lab Materials needed for the unit are listed together on one page for ease in gathering the materials and equipment (see page 7). At our house, we put all of the standard lab equipment in one sturdy box, to carry it to our work surface and to simplify clean up. Then we gather all the lab materials at the beginning of the course and put them in a separate box.

The list of "Lab Sheets to Copy" is after the "Lab Materials Needed" list on page 8.

Lesson Plans & Resources

Kym's Konstruction Kit and lesson plans begin on page 78. Daily instructions are given under each topic. You will need resources: either library books, encyclopedias or textbooks. Usborne's Science & Experiments books *The World of the Microscope* or the *Complete Book of the Microscope* are highly recommended. Both are easy for the children to read and understand. Perhaps use it for an introduction to a topic, then research in more depth with other resources. The books listed are the ones we enjoyed the most. Listed from easiest to hardest reading level, they cover most of the topics. Also, when at the library, look for other books near the same place on the shelves where these books are located.

Lab Notebook
Set up a lab notebook in a 3-ring binder. Use dividers labeled as follows:
① *Lab Sheets,* ② *Lab Sheets* ✓ (completed labs), ③ *Vocabulary* and ④ *Report.*

① Lab Sheets
There are 35 Lab Sheets in this unit. Each one corresponds with a lesson studied. Lab Sheet 35 (page 74) is to be used with many of the other labs, and will be noted in the instructions. Familiarize the student with this sheet in the following ways: using the terms, how to fill in the table, drawing what they observe in the microscope (you may use a grid on the circles), labelling the specimen and writing notes from their observations. **Prepared Lab Sheets** are found in the second half of this Microscope Unit beginning on page 38.
If you teach the student how to set up, use and clean up the microscope, you will then enable him to complete the rest of the labs with minimal assistance from you. The labs are set up with each step listed and a box to check off when each step is completed. Use Lab Sheet 3 to instruct in set up, use and clean up.

② Lab Sheets ✓
When a lab sheet is completed, it will go into the *Lab Sheets* ✓ section.

③ Vocabulary
Vocabulary words are in order by topic, on pages 75-76. Work on these words daily to give mastery. Playing games helps with learning and memorization, such as match the meaning, or concentration. You might type the words into your computer spelling program and have the student practice them on the computer. Or write and define them, putting them in the *Vocabulary* notebook section. *Just work with the vocabulary words daily to give mastery.* **There is a reminder on the lesson plans to check them off daily.**

④ Research Report
There is minimal writing in this unit since the focus is on learning to use the microscope and report on the labs. Written with my hands-on, no-write son in mind, this Microscope Adventure! Unit should be easy for children from fourth grade and up to accomplish. The *Report* section is for the biographical or topical report the student will research and write. One research report is suggested, with the length determined by the teacher. There are two options for completing the report, with numerous variations:

1. Work on the report for short periods of time daily for the duration of the unit.
2. Spend several days working on the report, investing longer periods of daily time.

Websites
Internet websites are listed on page 90. Please read the caution about using the Internet, and letting students use it alone. These sites were "there" and "clean" as of spring 2005. Also, check the list for topical websites to use in your daily teaching.

Ending the Unit
To end the unit, you might want to make a book or folder from the Lab Sheets, observations and the report the student writes. On page 36 is a suggested ending method for the unit.

Lab Materials Needed

When a lab lists "Get lab equipment," these are the standard supplies:

- ❑ pencil
- ❑ lab report sheet
- ❑ microscope
- ❑ slides
- ❑ cover slips
- ❑ pipette (eyedropper)
- ❑ forceps (tweezers)
- ❑ scalpel
- ❑ cutting board
- ❑ magnifying glass
- ❑ blotting paper (paper towel)
- ❑ purified water
- ❑ a flat working surface

Suggestion: put all the lab equipment in a ***box*** for ease of carrying, use, clean-up and storage.

Use Lab Sheet 35 for labs requiring a drawing of what you see through the microscope. The table is to be filled in: Date (9/19/2007), Specimen (insect leg), Mount (dry mount), Magnification (100x), Measurement (8 units or 2.5 mm), Measuring Tool (graticule, ruler or micrometer). Label each drawing with specimen name, matching number of drawing to number on the table.

One prepared slide is necessary, others may be added. Some additional prepared slides we found helpful were: human muscle (3 types), blood, woody and herbaceous stems, monocots and dicots, various microbes, diatoms, different types of fibers and exotic insects (whole or in parts).

Lab 3
prepared slide

Lab 4
specimen (food, dirt)

Lab 5
specimen (leaf, fingernail)
microtome (or cutting board)

Lab 6
specimen
mounting material (Euparal)
labels or
wax-tipped pencil
slide container

Lab 7
rubber or metal washer
waterproof glue
specimen
shredded paper towel or cotton

Lab 8
knit fabric
woven fabric
man-made fabric
natural fabric

Lab 9
toilet paper
paper towel
newspaper
stationery
other paper products
letter "e" from newspaper

Lab 10
human hair
various animal hairs
feather

Lab 11: Film
negatives
black & white photographs
color photographs
video film
audio film

Lab 12: Dust
dust from several places

Lab 13: Crystals
timer
$^1/_2$ teaspoon measuring spoon

salt
boric acid
sand
other crystal: sugar, brown sugar, abrasive cleanser

Lab 14: Bubbles
liquid dish soap
1 teaspoon milk

Lab 15: Cells
scissors or scalpel
cutting board
flowers

Lab 17: Flowers
flower or several flowers (with pollen)

Lab 18: Water Ducts
plant with roots, stem and leaves
clover
microtome

Lab 19: Pollen
flowers with pollen

Lab 20: Seeds
variety of seeds: maple, lentils, lima beans, corn

Lab 21: Spores & Moss
fern frond with spores
moss

Lab 22: Mold
citrus fruit
small bowl
larger bowl
shoe box
alcohol

Lab 23: Yeast
1/2 teaspoon of yeast
1/2 teaspoon of sugar
small container to mix in
timer
1 tablespoon warm water

Lab 24: Cork
piece of cork
hand microtome or scalpel

Lab 25: Wood
pencil shavings

Lab 26: Epithelial Cells
toothpick
stain, iodine or food coloring

Lab 27: Muscle Cells
uncooked steak, intestine and heart (ask your butcher)
methylene blue stain (or blue food coloring)
iodine (or red food coloring)

Lab 28: Blood Cells
lancet
timer
methylene blue stain (or blue food coloring, or iodine)
sterilized needle

Lab 29: Microbes
3 small jars with lids
3 different stagnant water samples: pond, puddle, dog dish, other
mud
plants, grass or hay
labels

Lab 30: Diatoms
diatomaceous earth (DE) (used for swimming pools)

Labs 31-34: Insects
grasshopper
fly
bee
ant
beetle
butterfly
moth
mosquito
other

Lab Sheets to Copy

One copy of each lab, unless otherwise listed:

Lab Sheet 13 - 4 copies.
Lab Sheet 14 - 2 copies.
Lab Sheet 17 - 1-3 copies.
Lab Sheet 31 - 5-8 copies.
Lab Sheet 32 - 1-2 copies.
Lab Sheet 35 - 25-30 copies.

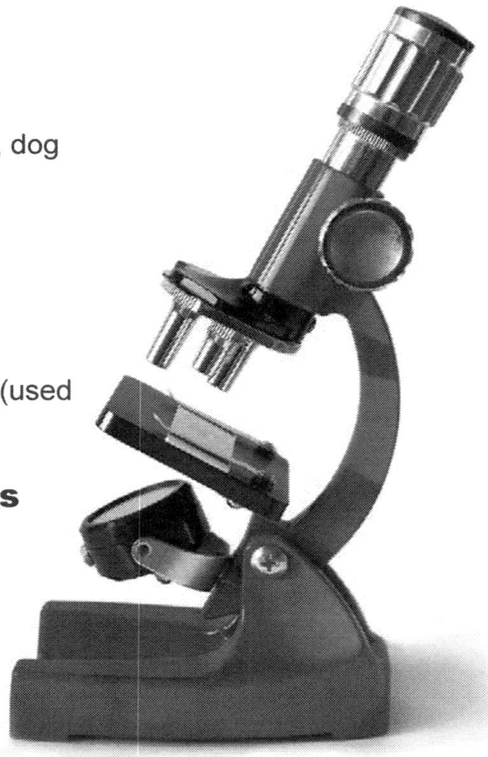

Books

Highly recommended: *The World of the Microscope* or the *Complete Book of the Microscope* by Usborne.

Microscope: Beginning to More Advanced

Greg's Microscope by Millicent E. Selsam - J 578 Sel
What's Smaller Than a Pygmy Shrew? by Robert E. Wells - J 539.1 Wel
The Microscope by Maxine Kumin - J 578 Kum
Microscope: How to Use It and Enjoy It by Eve and Albert Stwertka - J 578 Stw
The Microscope Book by Shar Levine - J 502.82 Lev
Experimenting with a Microscope by Maurice Bleifeld - J 598 Ble

General

Cell Wars by Dr. Fran Balkwill - J 616.079 Bal
Antoni van Leeuwenhoek: First to See Microscopic Life by Lisa Yount - JBiography

Molecules

Adventures with Atoms and Molecules by Robert C. Mebane - J 540.78 Meb
Molecules by Janice VanCleave - J 540.78 Van

Lenses

What's the difference between ... Lenses and Prisms and Other Scientific Things? by Gary Soucie - J 500 Sou
LENSES! Take a Closer Look by Siegfried Aust - J 681.4 Aus

Microscopic Life

MicroAliens: Dazzling Journey with an Electron Microscope by Howard Tomb - J 574 Tom
Hidden Worlds: Pictures of the Invisible by Seymour Simon - J 502.22 Sim
Extremely Wierd Micro Monsters by Sarah Lovett - J 576 Lov
YUCK! A Big Book of Little Horrors by Robert Snedden - J 579 Sne

Germs

Germs Make Me Sick! by Melvin Berger - J 616.92 Ber
Bacteria and Viruses (A New True Book) by Leslie LeMaster - J 616.01 Lem
Germs: Mysterious Microorganisms by Don Nardo - J 616.01 Nar
Viruses by Howard and Margery Facklam - J 576.64 Fac
Focus on Medicines by Susan DeStefano - J 615.1 Des

Inventions and Inventors

The Inventors: Nobel Prizes in Chemistry, Physics, and Medicine by Nathan Aaseng - J 609 Aas
The Industrial Revolution by Philip Wilkinson - J 609 Ing
Medical Technology: Inventing the Instruments by Robert Mulcahy - J 610.28 Mul
Invention by Eyewitness Books (Lionel Bender) - J 609 Ben
How Science Works by Judith Hann - J 507.8 Han
The History of Medicine by Steve Parker - J 610.9 Par
Through the Microscope: Frontiers of Medicine by Lionel Bender - J 610.9
Great Discoveries and Inventions by David Lambert - J 608.7 Lam

Microscopes

I. Parts of a Microscope

A. Read about the parts of a *microscope*, and study a picture of the parts.

B. Learn to identify the parts, especially the vocabulary words.

C. Read what each part is used for or its function.

D. Answer and do the following:

1. Make a list of the different types of *microscopes*.
2. Briefly explain the uses of the different types of *microscopes*.
3. What type *microscope* do you have?
4. Compare a *microscope* with a *magnifying glass*.
5. Contrast a *microscope* and a *magnifying glass.*

E. Compete Lab Sheet 1, labelling the parts of the microscope.

II. Stage & Lighting

A. Read about the microscope *lighting sources.*

B. Answer the following questions:

1. What are the two main *light sources* for a microscope?
2. Which type *light source* does your microscope have?
3. What is *top lighting*?

C. Read about the *microscope mirror* and *convex lenses.*

D. Answer the following questions:

1. What is the *mirror* used for?
2. Does the *mirror* move or is it a stationery fixture?

Microscopes

3. Are both sides of the *mirror* (top and bottom) the same?
4. What does *convex* mean?
5. What is a *convex lens*?
6. Which side of the *mirror* should you use for the *high-power lens*?

E. Read about the microscope *diaphragm.*

F. Answer the following questions:

1. Where is the *diaphragm* located?
2. What is the purpose of the *diaphragm*?

III. Objective Lenses & Magnification

A. Read about *objective lenses* and *magnification.*

B. Answer and do the following:

1. What does *low-power objective* mean?
2. What does *high-power objective* mean?
3. Is there a *medium-power objective*?
4. What do the numbers on the side of an *objective lens* mean?
5. How do you figure the magnification of an *objective lens*?
6. Where are the *objective lenses* mounted?
7. To what are they attached?
8. Is the object in #7 a set fixture or does it move?
9. Which *objective lens* is longest?
10. Which *objective lens* is shortest?
11. Read about and define *removable objectives*.
12. What is an *oil immersion objective* and how does it work?
13. How is an *oil immersion objective* different from regular *objectives*?
14. What do they usually study with *oil immersion objectives*?

C. Complete Lab Sheet 2.

History of Microscopes

History of the Microscope

A. Read about the history of microscopes.

B. Answer and do the following:

1. What did the first *microscopes* look like? Illustrate your answer.
2. Who was *Antony van Leeuwenhoek*?
3. Did he *invent* the *microscope*?
4. What role did he play with the *microscope*?
5. How do *microscopes* work? Briefly explain.
6. What does *light refraction* have to do with the microscope?
7. How were eyeglasses invented?
8. Who is *Hans Lippershey*?
9. Who is *Zacharias Janssen*?
10. Who is *Francesco Stelluti*?
11. Who is *Robert Hooke*?
12. What is *Micrographia*?
13. Who is it by?
14. What are *"animalcules"*?

C. Write a report on one of the people influential in developing microscopes. See list of "People to Study" on page 5. Work on the report daily in addition to lessons and labs. Consult teacher about length of report:

- ❑ 1 paragraph
- ❑ 2 paragraphs
- ❑ 3 paragraphs
- ❑ 1 page (750 words, typed)
- ❑ 2 pages (1500 words, typed)
- ❑ 3-5 pages, typed
- ❑ other

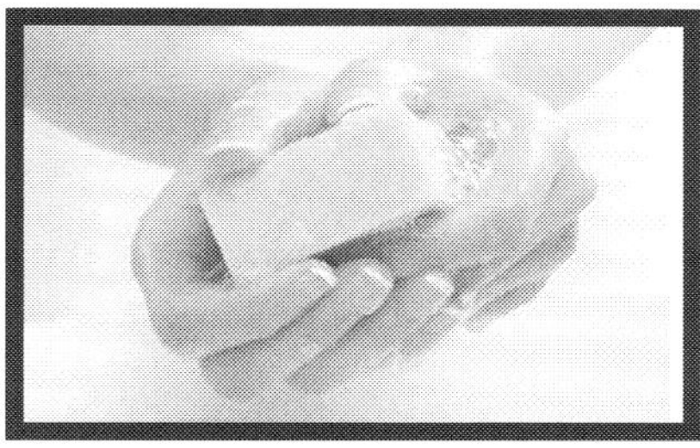

Set-Up, Use & Clean-Up

I. Setting up the Microscope

A. Read about setting up the *microscope*.

B. Read the cautions about carrying the microscope.

C. Read the cautions about using *direct sunlight* as a *light source*.

D. Answer the following questions:

1. Why do you view the microscope from the side as you lower the *objective*?
2. Which *power objective* do you begin with?
3. Why do you not use *direct sunlight* as your *light source*?
4. Can the light source ever be too bright?
5. Does cutting down on the light ever help make the image clearer?

E. Using Lab Sheet 3, practice setting up the microscope.

Keep Lab Sheet 3 available as a handy reference in future labs.

II. Using the Microscope

A. Read about moving the slide on the stage.

B. Answer the following questions:

1. When you move the *slide* to the right, which way does the *image* move?
2. When you move the *slide* to the left, which way does the *image* move?
3. Which way does the *image* move when you move the *slide* up?
4. Which way does the *image* move when you move the *slide* down?
5. What is *stage focusing*?
6. Does your microscope have *stage focusing*?

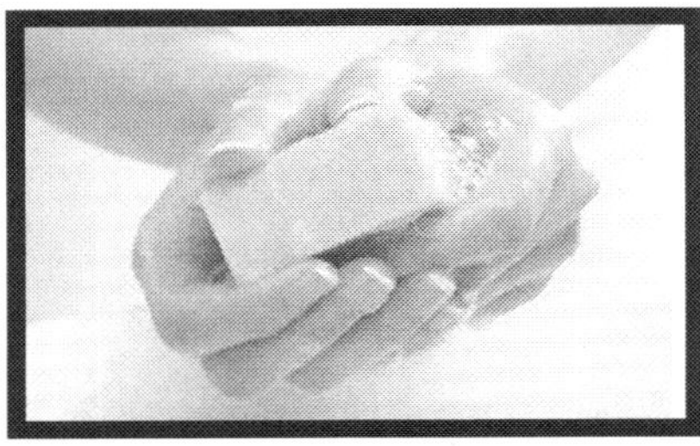

Set-Up, Use & Clean-Up

C. When working with specimens: ***never touch your face, eyes, nose, mouth - or anyone else's.***

D. Answer these questions about using the *high-power objective*.

1. Why do you look from the side to change the lens setting?
2. Why do you open the *diaphragm* more with the *high-power objective?*
3. Why do you use the *fine adjustment knob*, rather than the *coarse adjustment knob?*

III. Drawing Observations

A. Read about drawing what you see through the microscope and using *grids* to draw.

B. Using Lab Sheet 3 and a prepared slide, practice using the microscope. Use low-power first, then high-power.

C. Using Lab Sheet 35, draw what you see on *low-power*.

D. Label and fill in the table (figure magnification).

E. Using a prepared slide, practice using the *high-power objective*.

Remember to look from the side to change lenses.

F. Using Lab Sheet 35, draw what you see on *high-power*.

G. Label and fill in the table (figure magnification).

H. Make a list of cautions for using the microscope.

I. Make a poster from the list above. Hang it near your work area.

IV. Cleaning Up

A. Clean up your experiment, using Lab Sheet 3.

B. Keep it available as a handy reference in future labs.

Lab Equipment & Technique

I. Types of Slides

A. Read about the types of *slides*.

B. Answer and do the following:

1. What is a *microscope slide*?
2. Which different materials are used for making *slides*?
3. List the varying thicknesses available in slides.
4. List uses for the various thicknesses.
5. What is a *cover slip*?
6. What material are they made from?
7. What is a *stage micrometer*?
8. What is a *specimen*?

II. Slide Mounts

A. Read about the different *slide mounts*.

B. Answer the following questions:

1. What does *mount* mean?
2. What is a *temporary mount*?
3. What is a *permanent mount*?
4. What is a *prepared slide*?
5. What is a *dry mount*?
6. What is a *wet mount*?
7. What is a *semipermanent slide*?
8. What are *mounting agents* (or *mounting materials*)?
9. Make a list of the various *mounting agents*.
10. Which *mounting agent* is most recommended?
11. What is *Euparal*?

Lab Equipment & Technique

C. Using Lab Sheets 4-7, practice making each type of mount: *wet mount, dry mount, permanent mount* and *well mount.*

III. Other Lab Equipment

A. Read about lab equipment.

B. Answer these questions:

1. What is a *graticule*?
2. What is a *slide box*?
3. What is a *petri dish*?
4. What is a *petri dish* used for?
5. What is a *crystallizing dish*?
6. What is a *watch glass*?
7. What is a *pipette*?
8. What is the common name for *pipette*?
9. List various instruments used to cut specimens.
10. What are *forceps*?
11. What is the common name for *forceps*?

IV. Microtome

A. Read about *sections, cutting specimens, hand microtomes* and *microtomes*.

B. Answer the following:

1. What is a *microtome*?
2. What is a *section*?
3. How thin can a *microtome* slice a *specimen*?

C. Find instructions in a reference book and make a *hand microtome.*

Lab Equipment & Technique

D. Practice slicing specimens, making sections of various items (carrots, onions, plants) with your *hand microtome*.

E. Make slides of your *sections*.

F. View slides under the microscope.

G. Use Lab Sheet 35 to draw what you see through the microscope.

H. Label your drawings. Fill in the table.

Clean up all equipment. Wash everything accordingly.

Fibers

I. Fibers

A. Read about *fibers* and fabrics. Study how various fabrics are made, especially *knit* and *woven*.

B. Read about animal fur, yarn and thread.

C Answer and do the following:

1. What is a *fiber*?
2. What is a *natural fiber*?
3. What is a *man-made fiber*?
4. Is yarn made from one fiber or many fibers?
5. What gives fabric its pattern?
6. Make a list of *natural fibers*.

D. Complete Lab 8 with fabric, yarn and fibers.

E. Use Lab Sheet 35 to draw your observations.

F. Label and fill in the table.

II. Paper

A. Read about the process of making various types of paper: toilet paper, paper towels, newspaper, stationary, writing paper, paper plates and others.

B. Complete Lab 9 with various types of paper products. Cut the letter "e" from a newspaper to view.

C. Use Lab Sheet 35 to draw your observations.

D. Label and fill in the table.

III. Skin Covering

A. Read about human hair, animal hair and fur and feathers

B. Complete Lab 10 with various types of fibers: feathers, human hairs (head, arm, leg), animal hairs (gather as many as possible, ask neighbors and friends - dog, cat, sheep, goat, horse, hamster ...).

C. Use Lab Sheet 35 to draw your observations. Label and fill in the table.

D. Additional Labs: view fish or snake scales under the microscope. Draw and label.

Film & Dust

I. Film

A. Read about film, negatives, photographs, slides, black and white photos, color photos, video tape and audio tape.

B. Read about *top lighting*.

C. Answer these questions:

1. What is *top lighting*?
2. When would you use *top lighting*?
3. When would you use *bottom lighting*?
4. Where do you aim the *light source*?
5. What type of light fixture can you use for *top lighting*? (Get permission from an adult to move and use a light fixture.)
6. How much *reflected light* do you need from the mirror?

D. Collect as many different types of film, tape and photographs as you can find.

E. Complete Lab Sheet 11.

II. Dust

A. Read about dust.

B. What is dust made of?

C. Scrape a small amount of dust onto a slide and complete Lab Sheet 12.

D. Complete Lab Sheet 12 with dust from different locations: under the refrigerator, on top of the refrigerator, under the bed, near an animal's bed, near an open window.

E. Label the location and draw what you see.

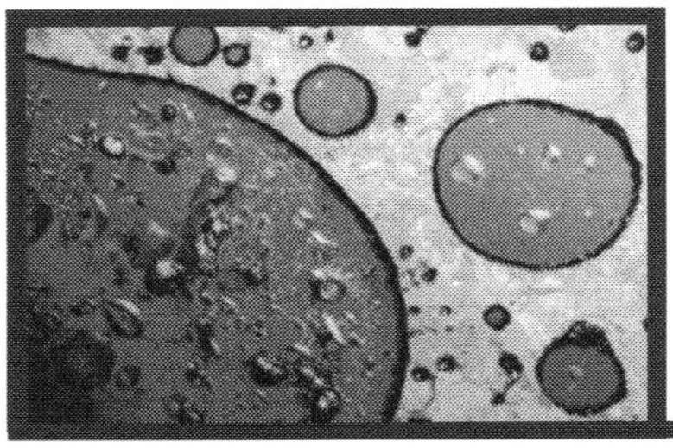

Crystals & Bubbles

I. Crystals

A. Read about *crystals, crystal formation* and *shape*.

B. Gather resources with pictures of *microscopic crystals*.

C. Answer the following questions:

1. What is a *crystal*?
2. How are *crystals* formed?
3. What can you think of that is made with *crystals*?
4. Are crystals *two-dimensional* or *three-dimensional*?
5. Do all *crystals* have the same shape?
6. What is sand made of?

D. Complete Lab Sheet 13.

E. Repeat Lab Sheet 13 with various other crystal substances.

II. Bubbles

A. Read about bubbles: how they are formed, how they appear under a microscope and how to prevent them when mounting a slide.

B. Answer these questions:

1. What is a *bubble*?
2. How are *bubbles* formed?
3. How do they appear under a microscope?
4. How can you prevent a bubble from forming when making a wet mount slide?
5. What are *globules*?

C. Complete Lab Sheet 14.

D. Complete Lab Sheet 14 again with various other liquids.

Stains & Dyes

Stains & Dyes

A. Read about *stains* and *dyes*.

B. Answer and do the following:

1. What is a *stain*?
2. What is a *dye*?
3. Is there a difference between a *stain* and a *dye*?
4. What is *iodine* used for?
5. What is *methylene blue*?
6. What is it used for?
7. What is *natural red*?
8. What is it used for?
9. What is *Carmine*?
10. What is it used for?
11. What is *acid fuchsin*?
12. What is it used for?
13. What is *Eosin*?
14. What is it used for?
15. When working with stains and dyes, what is *isopropyl alcohol* used for?
16. Make a list of stains used in *microscopy*.
17. Make a list of dyes used in *microscopy*.
18. Make a list of the various *staining techniques.*
19. What is *dipping and rinsing*?
20. What is *pulling stains*?
21. What is a *smear*?
22. What is *staining smears*?
23. Which technique is best for staining solid specimens?
24. Which technique is best for staining a *cell* to show the *nucleus*?
25. What are *color (colour) filters*?
26. What are they made of?
27. How are they used in *microscopy*?

Stains & Dyes

C. In future labs, you will use stains and dyes. Spend one lab practicing each of the stains and techniques listed in B.18 on page 21.

D. Using Lab Sheet 35, draw what you see before adding the stain.

E. Draw what you see after adding the stain.

F. Label the drawings.

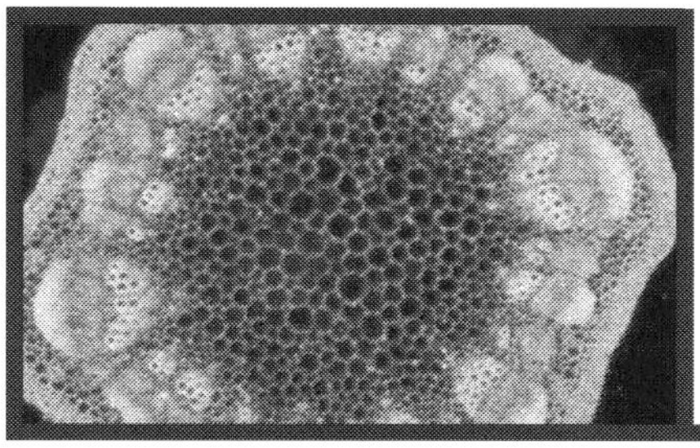

Cells & Plant Cells

I. Cells

A. Read about *cells, animal cells, plant cells, cell parts* and *function*.

B. Answer and do the following:

1. What is a *cell*?
2. List the parts of an *animal cell*.
3. List the parts of a *plant cell*.
4. Are there any differences between *plant cells* and *animal cells*?
5. If so, list them.
6. What is the *cell membrane*?
7. What is *cytoplasm*?
8. What is a *chromosome*?
9. What are *chloroplasts*?
10. What are *vacuoles*?
11. What are *granules*?
12. What is the *cell wall*?

C. Learn the parts of the cells.

D. Complete Lab Sheet 15.

II. Plant Cells: Overview

A. Read about *plant parts* and their *function*.

B. List the major parts of a plant.

C. Complete Lab Sheet 16.

III. Flowers

A. Read about flowers.

B. Answer these questions:

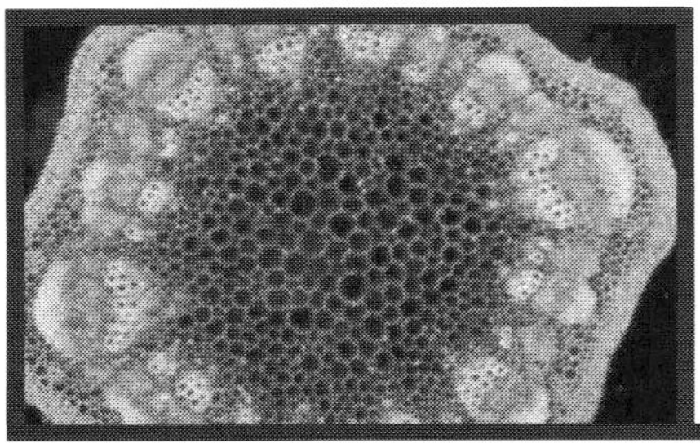

Cells & Plant Cells

1. What is a *flower*?
2. What is a *petal*?
3. What is the *ovule*?
4. What is *nectar*?

C. Complete Lab Sheet 17.

IV. Water Ducts

A. Read about *roots, root hairs, stalks, capillary action, chlorophyll, herbaceous stems, veins,* and *water absorption.*

B. Answer the following questions:

1. What is *capillary action*?
2. What is a *filament*?
3. What is a *leaf*?
4. What is a *vein*?
5. What is the *stalk*?
6. What is the *stem*?
7. What is *xylem*?
8. What is *phloem*?
9. What is *chlorophyll*?

C. Complete Lab Sheet 18.

V. Pollen

A. Read about pollen, the function of *pollen, stamen, pistil* and *anther*.

B. Gather and look at pictures of pollen magnified by a microscope.

C. Answer the following:

1. What is *pollen*?
2. Do all *pollens* have the same shape?
3. What makes *pollen* cling to other surfaces?

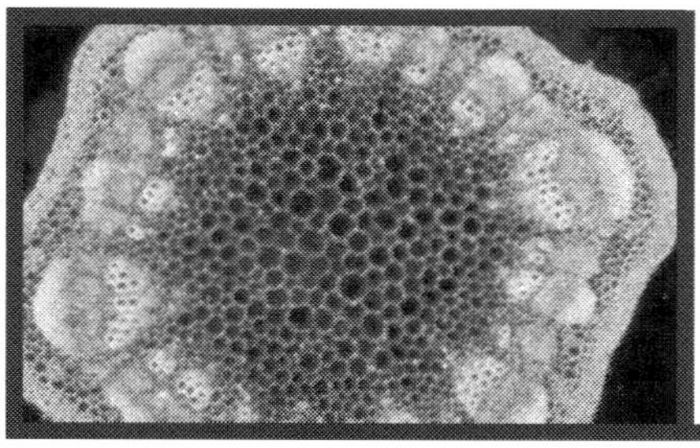

Cells & Plant Cells

4. What is the *anther*?
5. What is the *pistil*?
6. What is a *stamen*?

D. Complete Lab Sheet 19.

VI. Seeds & Fruit

A. Read about *seeds, seed parts* and their *function,* and *fruit.*

B. Look at various *seeds* or pictures of *seeds.*

C. Answer the following:

1. What is a *seed*?
2. What is its function?
3. Where does *fruit* develop on a plant?

D. Complete Lab Sheet 20.

E. Additional Labs: look at various fruit under the microscope. Use Lab Sheet 35 to record your observations.

Other Plants

I. Ferns & Spores

A. Read about *ferns, mosses* and *spores*.

B. Answer these questions:

1. What is a *spore*?
2. On which type of plants are *spores* found?
3. Where are they found on the plant?
4. What is a *spore case*?
5. What is a *fern*?
6. What is a *frond*?

C. Complete Lab Sheet 21.

II. Fungus, Mold & Mildew

A. Read about *fungus, mold* and *mildew*.

B. Answer the following questions:

1. What is *chlorophyll*?
2. Do *fungi* produce it?
3. What is the difference between the words *fungi* and *fungus*?
4. Make a list of fungi.
5. What is *mildew*?
6. What is *moss*?
7. What is *mold*?
8. What is an *antibiotic*?
9. What are antibiotics made from?
10. What is a *colony*?
11. What conditions help fungus, mold and mildew thrive?

C. Complete Lab Sheet 22.

Other Plants

III. Yeast

A. Read about *yeast*.

B. Answer and do the following:

1. Y*east* belongs to which plant group?
2. How many *cells* form a *yeast molecule*?
3. How do we use *yeast* in cooking?
4. Explain how *yeast* works in cooking.
5. What is a *parent cell*?
6. How do *yeast cells reproduce*?
7. What type of food does *yeast* like best?

C. Complete Lab Sheet 23.

IV. Cork

A. Read about *cork*, its *formation* and *structure.*

B. Answer the following questions:

1. *Cork* is made of what type of cells?
2. Are they living or dead?

C. Complete Lab Sheet 24.

V. Wood

A. Read about *wood,* its *formation, structure, cells,* and *woody stems*.

B. Complete Lab Sheet 25.

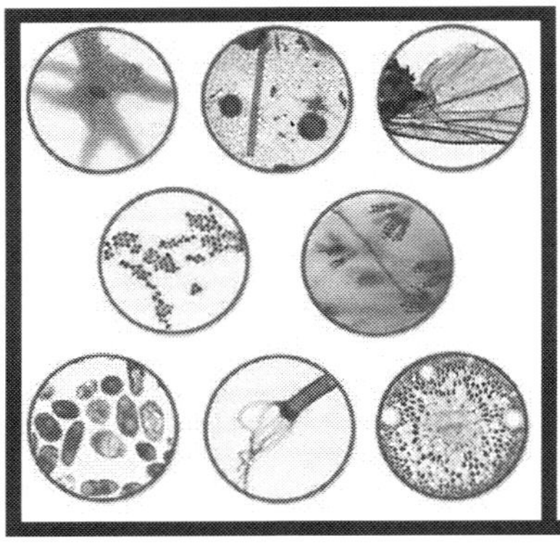

Animal Cells

I. Animal Cells

A. Read about *animal cells*.

B. Answer these questions (some questions are more in-depth than the earlier lesson):

1. What are *epithelial cells*?
2. What shape are *epithelial cells*?
3. Which tissues are made of *epithelial cells*?
4. What is a *nucleus*?
5. What shape is the *nucleus*?
6. What size is the *nucleus* in relation to other parts of a cell?
7. What are *chromosomes*?
8. What shape are *chromosomes*?
9. What is the function of *chromosomes*?
10. What is the *cytoplasm*?
11. What function does the *cytoplasm* serve?
12. What is the *cell membrane*?
13. What is the function of the *cell membrane*?
14. Are cells *two-dimensional* or *three-dimensional* in shape?

C. Complete Lab Sheet 26.

II. Muscle Cells

A. Read about *muscle cells*.

B. Answer and do the following:

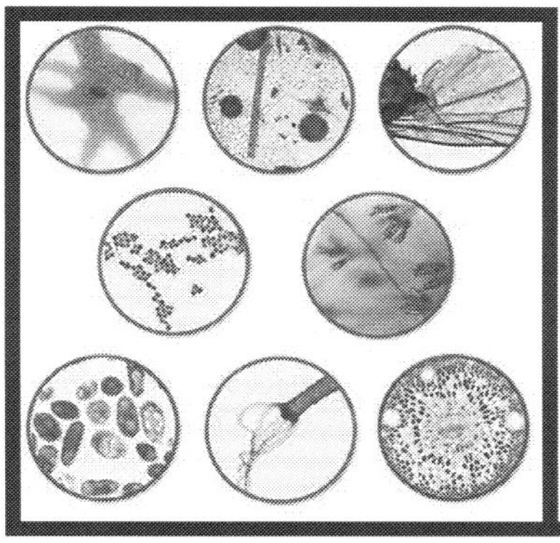

Animal Cells

1. What is a *voluntary muscle*?
2. Make a list of some of the *voluntary muscles* in the body.
3. What is an *involuntary muscle*?
4 Make a list of some of the *involuntary muscles* in the body.
5. What are *striations* or *striated tissue*?
6. What is a *smooth muscle cell*?
7. Is there a difference between the cells of *voluntary* and *involuntary muscles*?
8. If so, list the differences.
9. Which is larger, a *striated tissue cell* or a *smooth muscle cell*?
10. What do *muscle cells* look like?
11. Does a *muscle cell* have a *nucleus*?
12. Does a *muscle cell* have more than one *nucleus*?
13. What is *cartilage*?

C. Complete Lab Sheet 27.

III. Blood Cells

A. Read about *blood cells*.

B. Answer these questions:

1. Before pricking your finger for a drop of blood, why do you *sterilize* the needle first?
2. What are *erythrocytes*?
3. What does *"erythro"* mean?
4. What is the shape of *red blood cells*?
5. How many *erythrocytes* are in 1 milliliter of blood?
6. What is the function of *red blood cells*?
7. What color are *red blood cells*?
8. Do *red blood cells* have a *nucleus*?

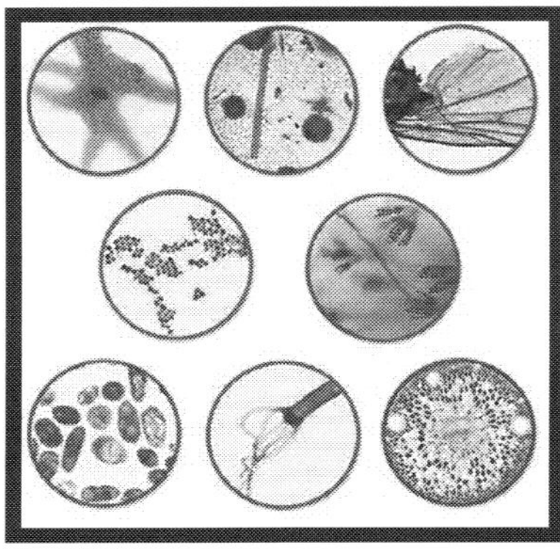

Animal Cells

9. What are *leukocytes*?
10. What color are *leukocytes*?
11. What are *white blood cells*?
12. Are there more *red blood cells* or *white blood cells* in a drop of blood?
13. What is the function of *white blood cells*?
14. What is the best *stain* to use in order to see *white blood cells*?
15. What is the best *stain* to use in order to see *red blood cells*?
16. Does the *white blood cell* have a *nucleus*?

C. Complete Lab Sheet 28.

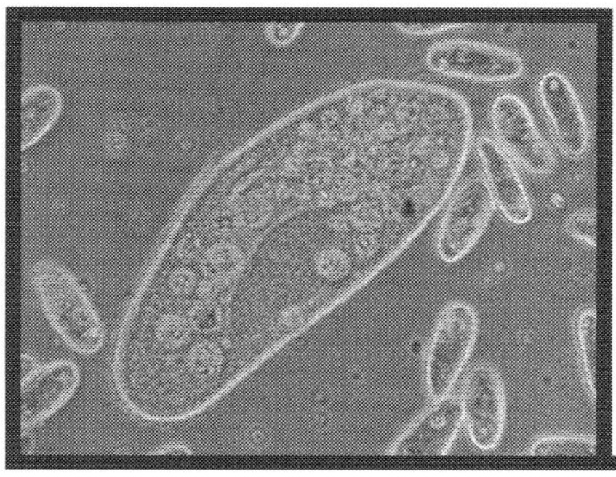

Microbes

I. Fresh Water Microbes

A. Read about *microorganisms*.

B. When doing the water experiments, try to get the book *Monerans & Protists* by Dr. Virginia Alvin and Robert Silverstein. In the back of that book is an "Identikey" to help you identify various microscopic organisms and classify what you are seeing.

C. Answer and do the following:

1. What is a *protozoa*?
2. How many different types of *protozoa* are there?
3. Make a list of the most common *protozoa*.
4. What are *Euglenas*?
5. What color are *Euglenas*?
6. Why?
7. What is a *flagellate*?
8. What is a *flagellum*?
9. What is a *pseudopod*?
10. Name some ways that *microorganisms* move.
11. What is a *paramecium*?
12. What is *cilia*?
13. What is *algae*?
14. What color is *algae*?
15. What are *desmids*?
16. What are *diatoms*?
17. What are *rotifers*?
18. Make a list of the most common *microscopic plant life* in water.

D. From books, draw pictures of the most *common microscopic life* in fresh water.

E. Label the pictures.

F. Complete Lab Sheet 29.

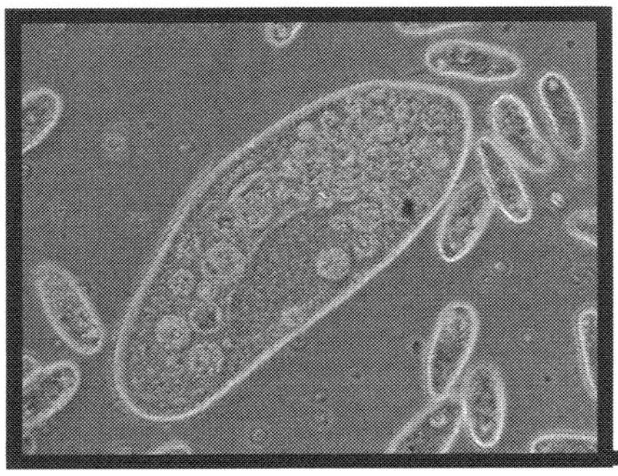

Microbes

II. Marine Microbes

A. Read about *microscopic life* in the ocean.

B. Gather resources containing pictures of *microscopic marine life*.

C. Answer and do the following:

1. What does *marine* mean?
2. Make a list of the main *microscopic marine animals*.
3. Is there a difference between the *microscopic marine animals* and the *microscopic animals* found in fresh water?
4. Make a list of the main *microscopic marine plants*.
5. Is there a difference between the *microscopic marine plants* and the *microscopic plants* found in fresh water?
6. What are *hydroids*?
7. What is a *polyp*?
8. What is a *crustacean*?
9. Can *crustaceans* be microscopic?
10. What is a *copepod*?
11. What is a *larvae*?

D. Complete Lab Sheet 30.

III. Soil Animals

A. Read about *microscopic animals* found in the soil.

B. What are the two basic categories of *microscopic animals* found in soil?

C. If you would like to further study *microscopic animals* found in soil, you may research methods and experiments to help you do this. (In the book *The World of the Microscope* by Usborne, you will find instructions and information to help you.)

D. Do an experiment to discover the *microscopic animal life* found in your soil.

Insects

I. Insects

A. Read about insects.

B. Answer and do the following:

1. What is an *insect*?
2. When using a microscope, why must we study most insects in parts rather than altogether?
3. Though we would rather find insect specimens that are already dead, sometimes it is necessary to anesthesize them to use as specimens. Considering this, what is the best way to kill an insect for observation?

II. Insect Legs

A. Read about insect legs.

B. Look at pictures of insect legs.

C. Answer and do the following:

1. Make a list of the most common insects in your area. *You can find insects around window sills, outside in the yard and near plants and flowers.*
2. How many legs do all *insects* have?
3. Describe the legs of a grasshopper. Are all the legs the same?
4. Do flying insects have powerful legs?
5. Can you tell about an insect's method of *locomotion* by its legs?

D. Gather the suggested insects and complete Lab Sheet 31.

III. Insect Antennae

A. Read about insect antennae.

B. Look at pictures of insect antennae.

Insects

C. Answer and do the following:

1. What is an *antennae*?
2. What is an *antennae* sometimes called?
3. Why?
4. *Antennae* are used as *receptors* for which senses?
5. Do all insect *antennae* look the same?
6. If there are variations, list them.
7. How many *antennae* does an insect have?
8. Insect antennae are attached to which body part?
9. Are insect *antennae* rigid or flexible?

D. Read about mosquitoes.

E. Answer the following:

1. Do male and female mosquitoes have identical *antennae*?
2. If there are differences in the antennae of the different sexes, list them.

F. Complete Lab Sheet 32.

IV. Insect Wings

A. Read about insect *wings*.

B. Answer and do the following:

1. What is a *wing*?
2. Do all insects have the same number of wings?
3. List the types of *coverings* found on insect wings.
4. What is special about the wing *coverings* of moths and some butterflies?
5. Can insect wings have *hair* on them?
6. Can insect wings have *scales* on them?
7. Are all insect wings *transparent*?

C. Complete Lab Sheet 33.

Insects

V. Insect Eyes

A. Read about insect *eyes*.

B. Answer and do the following:

1. How many *eyes* do insects have?
2. What type of eyes do insects have?
3. What is a *simple eye?*
4. What is a *compound eye*?
5. Do all insects have *simple* and *compound eyes*?
6. Are the eyes of insects located in the front of their head, like humans, or on the side of their heads like chickens?
7. Can insects move their eyes in their sockets?
8. While keeping their eyes still, can insects see only in front, or can they see all around?
9. How many images do insects see with their *compound eyes*?

C. Complete Lab Sheet 34.

VI. Insect Mouths

A. Read about insect mouths.

B. Answer and do the following:

1. What is a *proboscis*?
2. What are *mandibles*?
3. What type of food does an insect with a *proboscis* eat?
4. What type of food does an insect with *mandibles* eat?
5. What does the *proboscis* of a mosquito look like?
6. Is there a difference in the *proboscis* of a male mosquito and a female mosquito?
7. What does the *proboscis* of a butterfly look like?
8. Draw a picture of a butterfly's *proboscis*.

Ending the Unit & Field Trips

Ending the Unit

To end the unit, you might want to make a "book" from the Lab Sheets, observations and the report the student writes. Below is a suggested ending method, to bring the unit to a close:

1. ❑ Put the Lab Sheets in a paper or clear folder.
2. ❑ Make a cover picture for it.
3. ❑ Put the report in with the rest of the papers.
4. ❑ Make a Table of Contents.
5. ❑ Make a Title Page.
6. ❑ Staple the book together, complete with author's name on front.

If the microscope unit is used as part of a biology course, you might add these pages to your biology notebook.

Field Trips

You could also plan a field trip, or two. Ideas would include:

1. ❑ Visit a doctor's office.
2. ❑ Visit a medical technology laboratory.
3. ❑ Visit a blood donation center.
4. ❑ Visit a lens manufacturing factory.
5. ❑ Visit a science supply store.
6. ❑ Visit a school lab, either high school or college.
7. ❑ Visit a veterinarian's office.
8. ❑ Visit a science museum.
9. ❑ Visit a friend with a microscope.

Lab Sheets

NAME

DATE

Lab 1

Lab 1: Microscope Parts

1. ❑ Study a picture of a microscope, either online or in a book. Learn to identify the parts, especially the vocabulary words.
2. ❑ Using the picture below, label the parts of the microscope.
3. ❑ *Answers on page 77.*

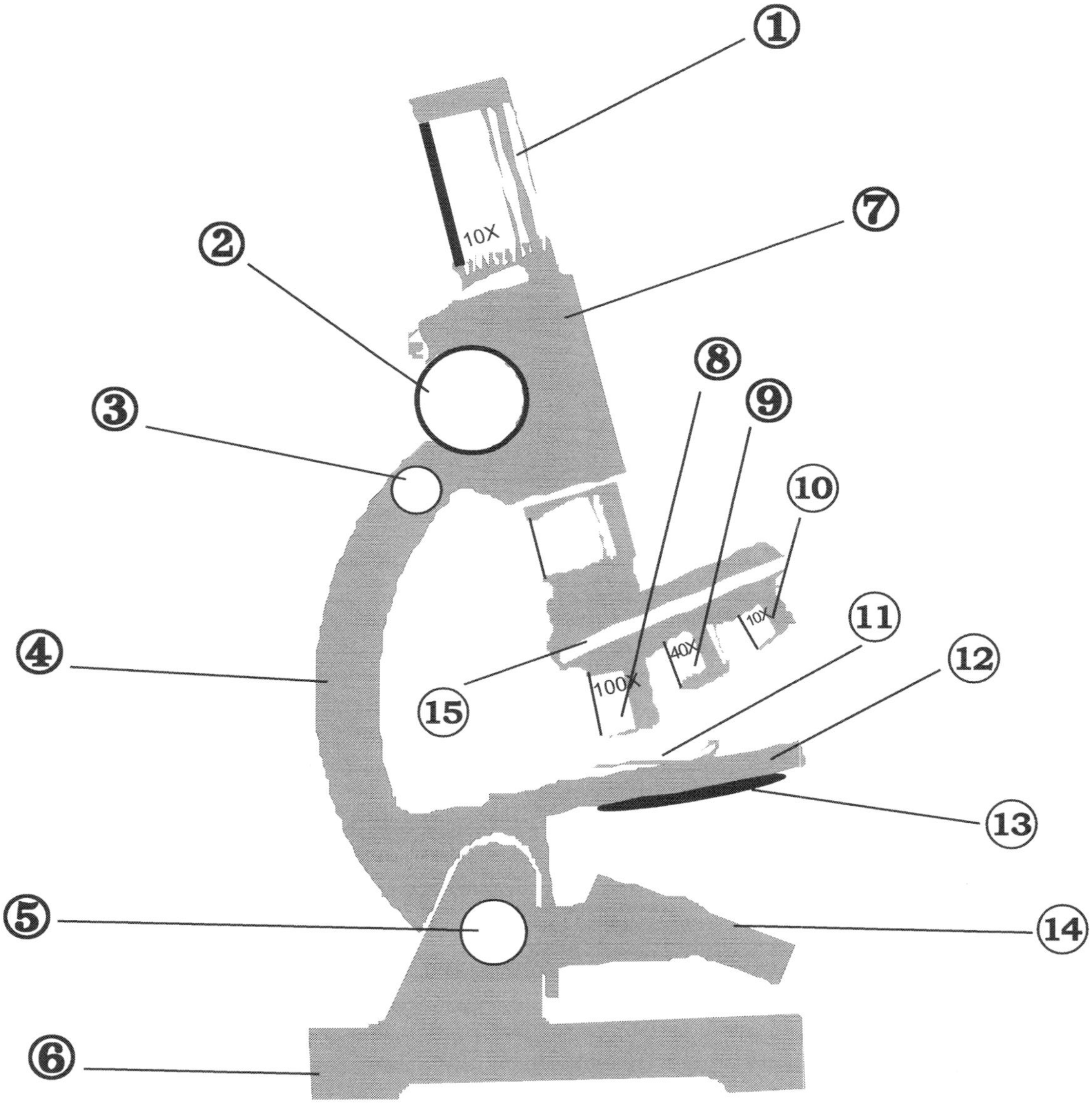

NAME

DATE

Lab 2

Lab 2: Magnification

1. ❑ Read about *magnification* of *specimens* with the *eyepiece* and the *objective lenses*.
2. ❑ Using the picture of a microscope from the previous lab, figure the *magnification* of the objectives: #8, #9 and #10. *Answers on page 77.*
3. ❑ ⑧
4. ❑ ⑨
5. ❑ ⑩
6. ❑ Look at your microscope and figure the magnification of each objective.

NAME ____________________
DATE ____________________

Lab 3

Lab 3: Set Up, Use & Clean Up

Caution: Never use *direct sunlight* as the light source for your microscope. It is too bright and might harm your eyes.

Materials Needed
microscope
prepared slide
table or flat surface for working

Setting Up the Microscope

1. ❑ If the microscope is in a case, take it out of the case by holding onto the arm.
2. ❑ Looking through the microscope, move the *convex mirror* into a position to give you good lighting or turn on the light or plug in the electrical cord.
3. ❑ Swing the stage clips to the sides, put the slide on the stage, and keep it in place by swinging the stage clips over the edges of the slide.

Using the Microscope

4. ❑ To view through the *low-power objective*, turn it until it clicks into place.
5. ❑ Looking from the side, watch the bottom of the low-power objective as you move it down to about $^1/_4$" above the slide. *(This keeps you from smashing the objective into the slide, and possibly damaging both.)*
6. ❑ Turn the coarse knob adjustment to *raise* the objective until the image comes into view.
7. ❑ While looking through the eyepiece, turn the coarse knob adjustment up or down a little to sharpen the image.
8. ❑ Turn the fine knob adjustment to make the image clearer.
9. ❑ While looking through the eyepiece, with your fingers under the stage, move the diagram back and forth to make the image clearer.

To use the *high-power objective*:

10. ❑ While looking through the *low power objective*, get the *image* as clear as possible.
11. ❑ Looking from the side, hold the *body tube* with one hand and turn the *high-power objective lens* on the *nosepiece* with the other hand. Make sure it clicks into place.
12. ❑ Rotate the *fine adjustment knob* either way, until the *image* looks clear and sharp.
13. ❑ You may need to open the diaphragm to admit more light.
14. ❑ Continue to use the *fine adjustment knob* to fine-tune, as needed.

Lab 3: Set Up, Use & Clean Up

continued

Lab 3

Cleaning up the Microscope and Experiments

15. ❑ Always make sure you clean up everything.
16. ❑ Do not touch the lenses with your fingers, so you won't get fingerprints or oil on them.
17. ❑ Use a special lens paper to lightly clean the lenses.
18. ❑ Clean up all specimens. Throw away whatever is not needed.
19. ❑ Wash all slides, slip covers and instruments used, in warm, soapy water.
20. ❑ If necessary, boil water and clean instruments, slides and slip covers in the hot water. *Adult supervision required. Be careful not to burn yourself.* Boil only when dealing with questionable specimens: insects, animal parts, feces, blood, etc.
21. ❑ Replace slides to their container.
22. ❑ Replace slip covers to their container.
23. ❑ Replace instruments to their proper place.
24. ❑ If you must save the specimen for later use, be sure it is stored properly and is out of reach of small children.
25. ❑ Cover the microscope with a dust cover or put it back in its case.
26. ❑ Put everything up. *Make sure it is out of reach of smaller children.*
27. ❑ Wash hands with warm, soapy water.

ANTIBACTERIAL SOAP

NAME ....

DATE ______________________________

Lab 4

Making Slides

Lab 4: Wet Mount

A slide (sometimes referred to as a "mount") is a thin, rectangular piece of glass on which a microscope sample is placed. To make a slide, one prepares the *specimen* and places it on the slide in one of the following ways:

wet mount	dry mount
permanent mount	well mount
temporary mount (may be wet or dry, but is kept for a short time)	

Follow these steps for making the various slides or mounts. See illustrations for materials needed.

A *wet mount* is a temporary way of preparing a microscope *specimen* in a liquid medium between the slide and cover slip. Usually the liquid medium is water, but it may be a stain or other liquid. Most biological specimens must be kept wet, or they could dry up and decay.

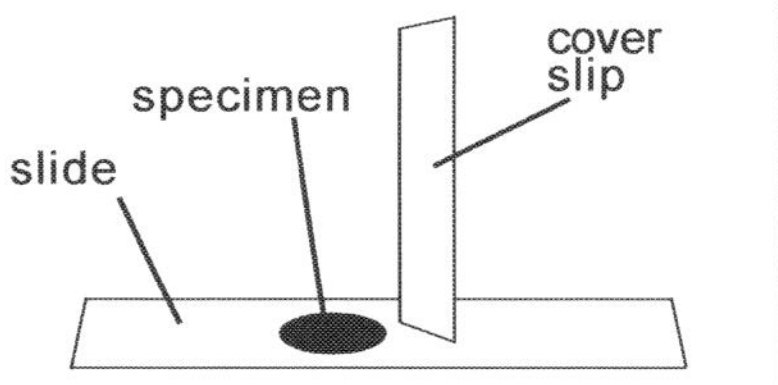

illustration A

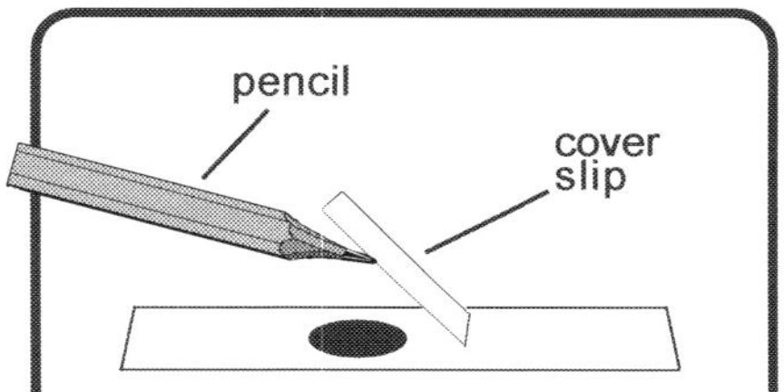

illustration B

1. ❑ With a pipette, put one drop of water in the center of a clean slide.
2. ❑ Using forceps, carefully place the specimen on top of the drop of water.
3. ❑ Holding the cover slip by its edges, stand it upright (at a 90° angle with the slide). See illustration A.
4. ❑ Move cover slip so one edge of the slip touches the edge of the drop of water.
5. ❑ Using a pencil or probe, gently lower the cover slip over the drop of water and the specimen, trying *not to trap any air bubbles.* See illustration B.
6. ❑ Soak up any excess water with a tissue, paper towel or blotting paper.
7. ❑ To wet a slide that is drying out: use your pipette to put a drop of water next to the cover slip on a side that still seems wet. The water will be pulled under the cover slip and will push out any air bubbles.
8. ❑ To save a wet mount for a longer time: use a toothpick to spread a thin layer of petroleum jelly around the outside edges of the cover slip. This acts as a seal to prevent evaporation.

Clean up all equipment. Wash everything accordingly.

NAME ____________________

DATE ____________________

Making Slides

Lab 5

Lab 5: Dry Mount

A dry mount is used for specimens that do not need rehydrating, like cork, wood and insect parts.

Materials Needed
clean slide
specimen
scalpel (or knife or scissors)
microtome (or cutting board)
microscope

1. ❑ Slice a thin section of the specimen.
2. ❑ Place this section on a clean microscope slide.
3. ❑ Usually a cover slip is not used since there is no liquid to "glue" the cover and slide together.
4. ❑ When working with a dry mount, use care when moving the microscope or the stage. The specimen might slip off the slide.

Clean up all equipment. Wash everything accordingly.

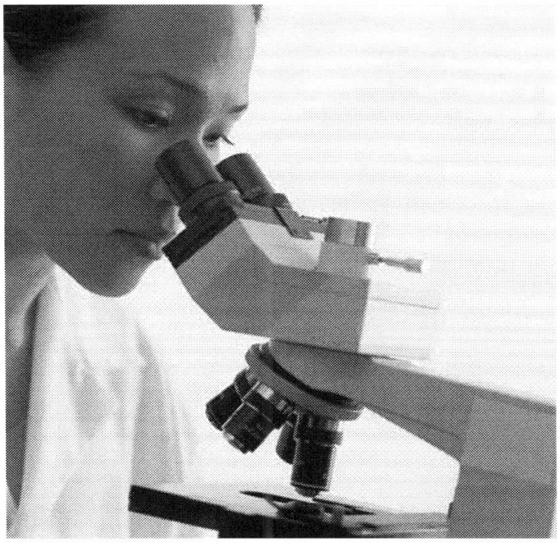

NAME

DATE

Making Slides

Lab 6

Lab 6: Permanent & Semi-Permanent Mounts

Materials Needed
clean slides
cover slips
mounting material
specimen
small labels *OR*
wax-tipped pencil
clear, white corn syrup

Permanent Mount

A permanent mount is a specimen preserved on a slide to view at a later time. Using the proper equipment and mounting materials (chemicals) and done properly, permanent mounts may last for many years.

1. ❑ Place specimen on a clean slide.
2. ❑ Put a few drops of mounting material on the specimen.
3. ❑ Place a cover slip on top of the specimen and mounting material.
4. ❑ Label your specimen:
   - ❑ Using small labels, record the name of the specimen, the date and the stain used (if there was one). Place the label on one end of the slide.
   - ❑ Or write the above information directly on the slide with a wax-tipped pencil.
5. ❑ Let harden in a warm, dry place for several hours, or overnight.
6. ❑ To organize your slides:
   - ❑ Number them and keep a record of the specimens.
   - ❑ Or you can keep them in alphabetical order.

Semi-Permanent Mount

A semi-permanent mount is a specimen preserved on a slide to view at a later time, however, it won't last forever.

7. ❑ Place specimen on a clean slide.
8. ❑ Put a drop of clear, white corn syrup on the specimen instead of water.
9. ❑ Add a cover slip and it is ready to use.

Clean up all equipment. Wash everything accordingly.

NAME ....

DATE ....

Making Slides

Lab 7

Lab 7: Well Mount

A well mount is helpful with live, moving specimens such as microscopic organisms. This mount contains, or holds in, the water like a microaquarium, allowing you to view live specimens.

Materials Needed
clean slide
water
pipette
small rubber or metal washer
waterproof glue
specimen
shredded paper towel or cotton
blotting paper (tissue or paper towel)

1. ❑ Using permanent glue, affix a clean washer to the center of a clean slide. Wipe up any spilled glue.
2. ❑ Allow glue to dry.
3. ❑ Using a pipette, place some water (may hold several drops) in the well formed in the center of the washer.

To slow down the fast swimmers, so you can focus on them and view them better:

4. ❑ Shred a tiny piece of paper towel or cotton fabric.
5. ❑ Place the fibers into the well of the slide, making a "nest" of the fibers.
6. ❑ Place one drop of water in the well.
7. ❑ Cover the well with a cover slip.

Remember to keep the microscope stage level so the water doesn't spill out.

Read more about making "well mounts" to learn more uses and cautions.

You can also buy premade well mounts. These are microscope slides with a concave area in the center.

Clean up all equipment. Wash everything accordingly.

NAME

DATE

Making Slides

Lab 8

Lab 8: Fibers

Materials Needed
lab equipment
knit fabric
woven fabric
man-made fabric
natural fabric

1. ❑ Get lab equipment and make a temporary mount of the knit fabric.
2. ❑ Focus specimen under the microscope.
3. ❑ Adjust the light.

Remember to lower the low-power objective with your eyes looking from the side to make sure you do not hit the stage with the lens.

4. ❑ Using Lab Sheet 35 draw a picture of what you see in the microscope.
5. ❑ Label the drawing.
6. ❑ Fill in the table for the knit fabric.
7. ❑ Repeat with remaining types of fabric.

   On your own paper, answer and do the following:

8. ❑ Why is only a section of the fabric visible under the microscope at one time?
9. ❑ What do the fibers look like?
10. ❑ Under the microscope, do the edges of the fibers look smooth or rough? Write some descriptive words for each fiber:

    knit: ____________________

    woven: ____________________

    man-made: ____________________

    natural: ____________________

Clean up all equipment. Wash everything accordingly.

NAME ....

DATE ....

Making Slides

Lab 9

Lab 9: Papers

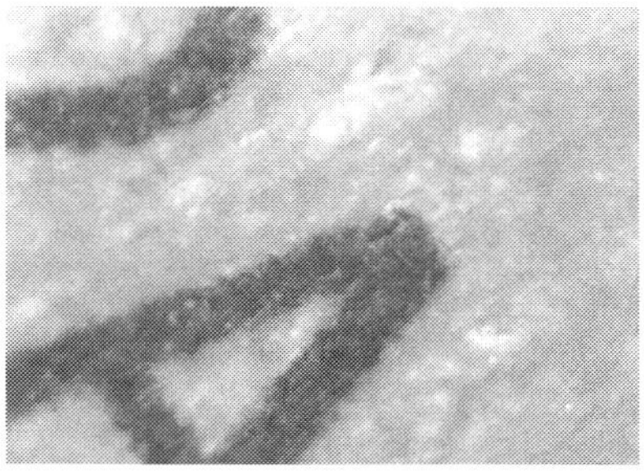

Materials Needed
lab equipment
letter "e" from newspaper
toilet paper
paper towel
newspaper
stationery
other paper products

1. ❑ Make a wet or semi-permanent mount of the letter "e" from the newspaper. *(If you make a semi-permanent mount, you can use this slide in future labs, to focus.)*
2. ❑ Get out lab equipment and focus specimen under the microscope on low-power.
3. ❑ Adjust the light.
4. ❑ Using Lab Sheet 35 draw a picture of what you see in the microscope.
5. ❑ Label the drawing.
6. ❑ Fill in the table for the paper.
7. ❑ Repeat with remaining types of paper.

   On your own paper, answer and do the following:

8. ❑ Why is only a section of the letter "e" visible under the microscope at one time?
9. ❑ Do all the paper fibers look alike? ❑ Yes ❑ No
10. ❑ Under the microscope, do the edges of the fibers look smooth or rough? Write some descriptive words for each fiber:

    letter "e": ______________________________

    toilet paper: ______________________________

    paper towel: ______________________________

    newspaper: ______________________________

    stationery: ______________________________

    other: ______________________________

    other: ______________________________

    other: ______________________________

Clean up all equipment. Wash everything accordingly.

NAME ....

DATE ....

Making Slides

Lab 10

Lab 10: Skin Covering

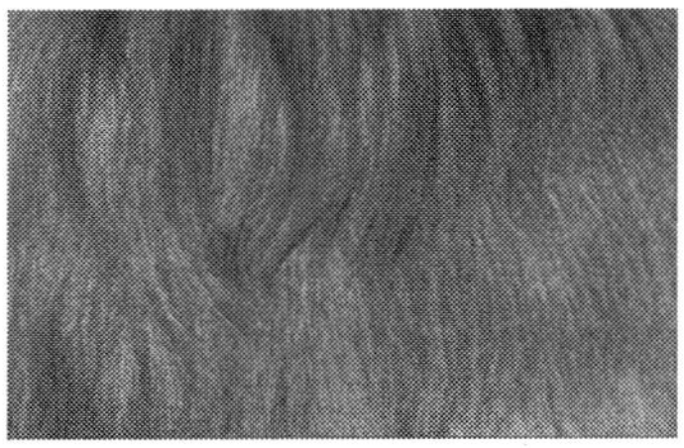

Materials Needed
lab equipment
human hair
various animal hairs
various types of feathers

1. ❑ Make a wet or permanent mount of a human hair.
2. ❑ Get out lab equipment and focus specimen under the microscope.
3. ❑ Adjust the light.
4. ❑ Using Lab Sheet 35 draw a picture of what you see in the microscope.
5. ❑ Label the drawing.
6. ❑ Fill in the table for the specimen.
7. ❑ Repeat with remaining types of hair and feathers.

   On your own paper, answer and do the following:

8. ❑ Do all the hairs look alike? ❑ Yes ❑ No
9. ❑ Can you see barbs on the edges of the feather?
10. ❑ Do you know what the barbs are for? (If not, you can research it.)
11. ❑ Under the microscope, do the edges of the hairs and feathers look smooth or rough? Write some descriptive words for each fiber:

    human hair: ____

    ________ hair: ____

    ________ hair: ____

    ________ hair: ____

    feather: ____

    feather: ____

    other: ____

    other: ____

    other: ____

Clean up all equipment. Wash everything accordingly.

NAME ______________________

DATE ______________________

Making Slides

Lab 11

Lab 11: Film

Materials Needed
lab equipment
negatives
black & white photographs
color photographs
video film
audio film
light source

1. ❑ Gather film and photographs. Get out lab equipment.
2. ❑ Get out lab equipment and light source.
3. ❑ Put the negative directly on the stage and clip in place.
4. ❑ Using *top lighting*, look at the different areas which make up the picture.
5. ❑ Do you see tiny dark dots? ❑ Yes ❑ No
6. ❑ On your own paper, explain what the dots are formed from.
7. ❑ Using Lab Sheet 35 draw what you observe. Label and fill in table.
8. ❑ Repeat this experiment with each of the other film types.
9. ❑ Using Lab Sheet 35 draw what you observe. Label and fill in table.
10. ❑ Repeat this experiment with the photographs.
11. ❑ Use the high-power objective.
12. ❑ Use good top-lighting.
13. ❑ Using Lab Sheet 35 draw what you observe through the microscope. Label and fill in table.

Clean up all equipment. Wash everything accordingly.

NAME

DATE

Making Slides

Lab 12

Lab 12: Dust

Materials Needed
lab equipment
dust from several sources

1. ❏ Get out lab equipment and scrape some dust onto a clean slide.
2. ❏ View the dust under the microscope on low-power.
3. ❏ Try to identify all the microscopic bits.
4. ❏ Using Lab Sheet 35 draw what you observe under the microscope.
5. ❏ Label what you can identify and fill in the table.
6. ❏ View the dust under the microscope on high-power.
7. ❏ Can you identify anything more clearly under high-power? ❏ Yes ❏ No
8. ❏ On the same Lab Sheet draw what you observe on high-power.
9. ❏ Label what you can identify and fill in the table.
10. ❏ Gather dust from the following places:
    - ❏ under the refrigerator
    - ❏ on top of the refrigerator
    - ❏ under the bed
    - ❏ near an open window
    - ❏ near an animal's bed
11. ❏ Draw what you observe under the microscope on low-power.
12. ❏ Label the location you found the dust, and what you can identify.
13. ❏ View the dust under the microscope on high-power.
14. ❏ Draw what you observe under the microscope on low-power.
15. ❏ Label the location you found the dust and what you can identify.

Clean up all equipment. Wash everything accordingly.

NAME ______________________

DATE ______________________

SUBSTANCE ______________________

Making Slides

Lab 13

Lab 13: Crystals

Materials Needed
lab equipment
timer
1/2 teaspoon (to measure with)
salt
boric acid
sand
other crystal(s)
(sugar, brown sugar, abrasive cleanser)

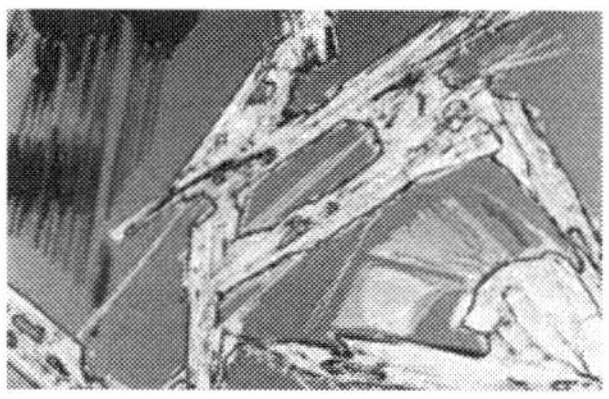

You will need at least 4 copies of this page.
Do this experiment first with salt.
Fill in all the blanks the first time through for the salt.
Repeat the experiment for each of the other substances, using a new sheet each time.

1. ❑ Get out lab equipment, timer, 1/2 teaspoon measure and water.
2. ❑ Get salt, boric acid, sand and other crystal substances.
3. ❑ Dissolve 1/2 teaspoon of ____________ (one substance per experiment) in 1 teaspoon of water. Stir.
4. ❑ Place a drop of the clear liquid on a clean slide and let it spread a little.
5. ❑ Put the slide, without a cover slip, aside for 3 minutes. Set the timer. Let it begin to dry.
6. ❑ Focus slide in the microscope on the outer portions of the drop.
7. ❑ Do you see crystals? ❑ Yes ❑ No
8. ❑ If so, what shape are they? On your own paper, describe the shapes in a few descriptive words.
9. ❑ Are they all the same size? ❑ Yes ❑ No
10. ❑ Move the light so that little or no light is coming up from the bottom. Can you see the crystals change to gleaming 3-dimensional shapes on a dark background?
11. ❑ On Lab Sheet 35 draw pictures of what you see in the microscope. Label and fill in table.
12. ❑ Do this same experiment with boric acid, then with sand, then with at least one other granular substance (sugar, brown sugar, or abrasive cleanser such as *Comet*). Answer the questions and draw pictures of what you see in the microscope.
13. ❑ Additional lab: you can repeat this experiment with the same substances, using a dry mount instead of a wet mount.

Clean up all equipment. Wash everything accordingly.

NAME ____________________

DATE ____________________

Lab 14

Making Slides

Lab 14: Bubbles

Materials Needed
lab equipment
liquid dish soap
1 teaspoon milk
pipette

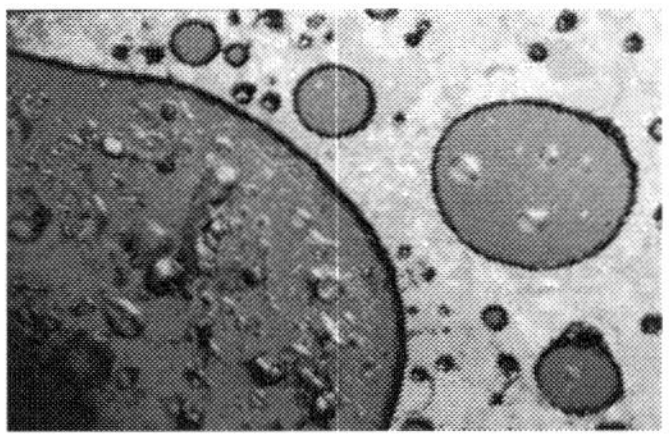

You will need at least 2 copies of this page. Do this experiment first with liquid dish soap and fill in all the blanks. Repeat the experiment for milk.

1. ❑ Get out lab equipment, liquid dish soap, and 1 teaspoon milk.
2. ❑ Place a small drop of liquid dish soap on a slide.
3. ❑ Using a second slide, push the drop lengthwise across the first slide to spread it.
4. ❑ Put a cover slip on.
5. ❑ Focus on the slide and move it back and forth. Do you see large and small circles with pearly centers and dark outlines? ❑ Yes ❑ No
6. ❑ Turn the mirror or light aside so no bottom light reaches the slide.
7. ❑ Does the background liquid look dark and the air bubbles seem like brightly gleaming rings? ❑ Yes ❑ No
8. ❑ On Lab Sheet 35 draw pictures of what you see on the soap slide through the microscope. Label and fill in the table.
9. ❑ Remove that slide and get out another slide.
10. ❑ This time make the slide with milk instead of soap. Repeat steps 2-8 with milk.
11. ❑ Do you see the air bubbles? ❑ Yes ❑ No
12. ❑ Do you also see small, round shapes floating in clear liquid? ❑ Yes ❑ No
13. ❑ These shapes are the globules of fat that give milk its white color.
14. ❑ Draw a picture of what you see in the microscope. Label and fill in table.
15. ❑ Additional Lab: do this experiment with various other liquids. Draw observations. Label and fill in table.

Clean up all equipment. Wash everything accordingly.

NAME ____________________

DATE ____________________

Lab 15

Lab 15: Cells

Study labeled pictures of plant and animal cells. Learn the parts. Decide which one is the plant cell, and which is the animal cell. Label the cells.

Label the parts. If a part is not shown, draw it in.

1. ❑ cell membrane
2. ❑ cell wall
3. ❑ cytoplasm
4. ❑ chromosome
5. ❑ chloroplast
6. ❑ vacuole
7. ❑ granule
8. ❑ nucleus
9. ❑ nuclear membrane
10. ❑ mitochondrion
11. ❑ chlorophyll

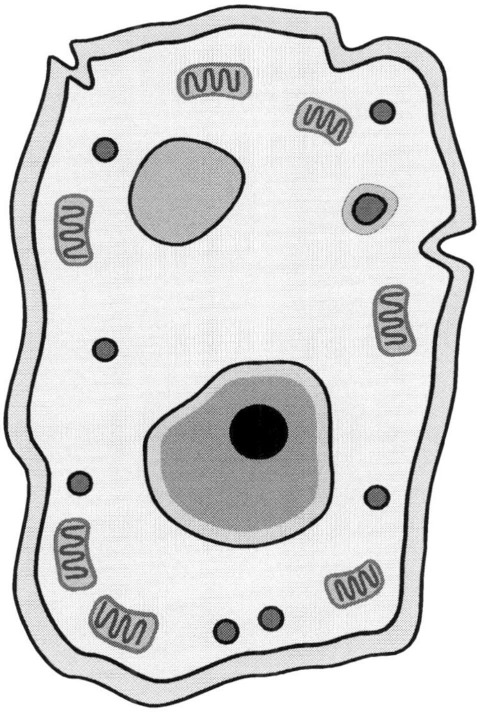

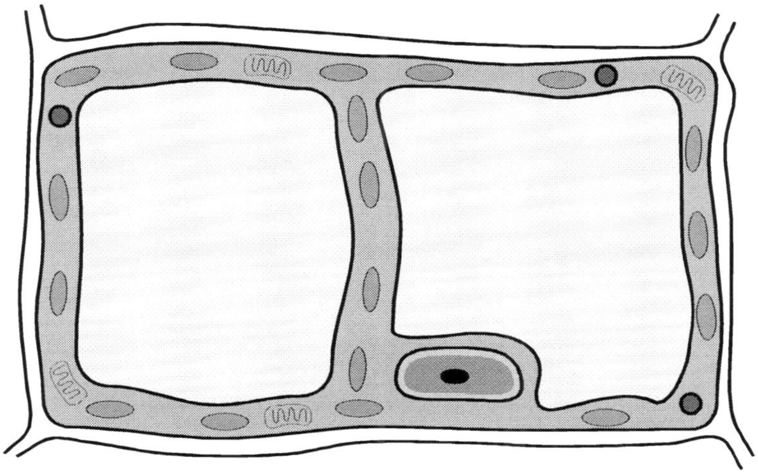

NAME

DATE

Lab 16

Lab 16: Plants

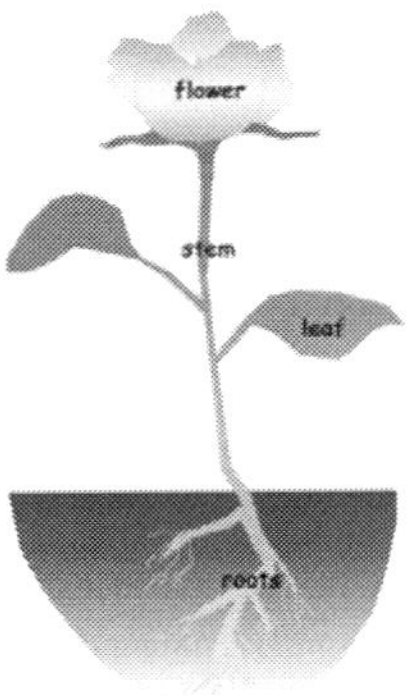

Draw a plant. Include and label the following parts:

1. ❑ flower
2. ❑ stem
3. ❑ leaves
4. ❑ root
5. ❑ root hair
6. ❑ anther
7. ❑ pistil
8. ❑ stamen
9. ❑ pollen
10. ❑ petal
11. ❑ ovule

NAME ______________________________

DATE ______________________________

Plant Cells

Lab 17

Lab 17: Flowers

Materials Needed
flowers
lab equipment

You might want to do this experiment with several flowers of different colors. If so, copy this page.

1. ❑ Get a bright-colored flower.
2. ❑ Get lab equipment.
3. ❑ Cut a small portion of a petal off the flower.
4. ❑ Mount petal piece between 2 **slides** (do not use a cover slip).
5. ❑ Look at it under the microscope (low-power).
6. ❑ Using Lab Sheet 35 draw what you see in the microscope.
7. ❑ Look at it under high-power.
8. ❑ Using Lab Sheet 35 draw what you see in the microscope.
9. ❑ Put the slide aside for 2 days.
10. ❑ Clean up equipment.

2 days later:

11. ❑ Get microscope and slide with petal.
12. ❑ View the petal under the microscope.
13. ❑ Draw a picture of what you see.
14. ❑ Is there any difference between the first drawing and the second? Briefly describe the difference: ______________________________
____
____
15. ❑ Briefly explain the difference: ______________________________
____
____
____

Throw away petal. Wash off slide and clean up equipment.

NAME ____________________

DATE ____________________

Plant Cells

Lab 18

Lab 18: Water Ducts

Materials Needed
lab equipment
clover
plant with:
roots
stem
leaves

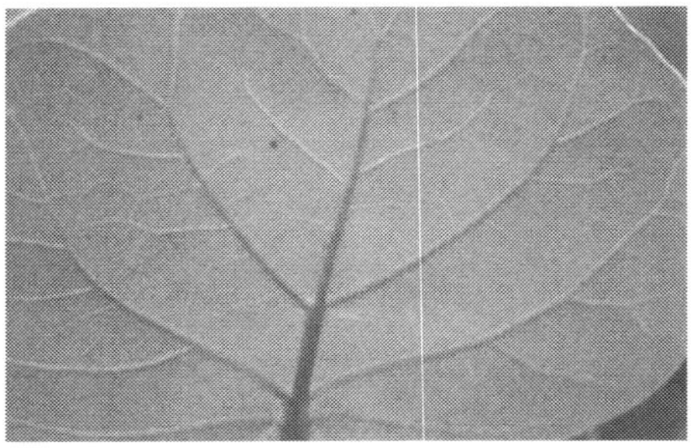

1. ❑ Get a clover and a plant with roots, stem and leaves.
2. ❑ Get lab equipment.
3. ❑ Do you see the tiny root hairs trailing from the roots? ❑ Yes ❑ No
4. ❑ Write a few descriptive words telling what they look like.

____

____

5. ❑ Using the scalpel & forceps on a cutting board, slice very thin cross sections of the stalk, roots, leaves and clover.

Always hold specimen with forceps (not your fingers) when using scalpel!

6. ❑ Mount each specimen on a separate slide without a cover slip.
7. ❑ Look at these under a microscope on low-power.
8. ❑ Can you see the channels in the specimen under microscope?
9. ❑ What is the name for these channels?
10. ❑ Can you locate a vein? (It looks like a hollow tube.)
11. ❑ Which tubes are the largest, the ones at the edge of the stalk, or the ones in the center?
12. ❑ Using Lab Sheet 35 draw what you see through the microscope.
13. ❑ Label and fill in the table.
14. ❑ Look at each one through the microscope on high-power.
15. ❑ Draw what you see through the microscope.
16. ❑ Label and fill in the table.

Clean up all equipment. Wash everything accordingly.

NAME

DATE

Lab 19

Plant Cells

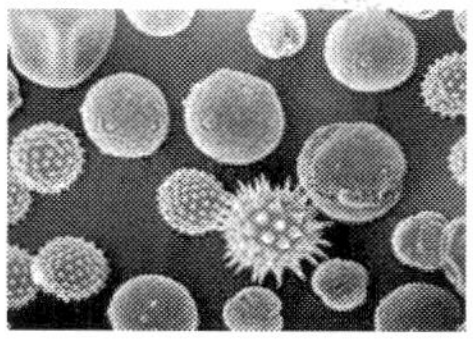

Lab 19: Pollen

You might want to do this experiment several times using different flowers with pollen. The pollen grains might look different in other plants.

Materials Needed
lab equipment
flowers

1. ❑ Get a flower or several flowers.
2. ❑ Get lab equipment.
3. ❑ Find the reproductive parts in the center of the flower. Locate the pollen.
4. ❑ Wipe some pollen onto a slide.
5. ❑ Add 1-2 drops of water and a cover slide.
6. ❑ Focus the pollen on low-power in the microscope.
7. ❑ Draw a picture of what you see. Label and fill in table.
8. ❑ Are the edges of the pollen smooth or rough?

   ❑ Smooth ❑ Rough
9. ❑ How many pieces of pollen do you see? (Write the number after the "#" sign, below.)
10. ❑ Wash off slide and clean up equipment.
11. ❑ Additional Lab: view the pollen on high-power. Draw what you see and label.

Pollen of: ____________ # ____________
plant name *# of pollen pieces*

Pollen of: ____________ # ____________
plant name *# of pollen pieces*

Pollen of: ____________ # ____________
plant name *# of pollen pieces*

Pollen of: ____________ # ____________
plant name *# of pollen pieces*

NAME

DATE

Plant Cells

Lab 20

Lab 20: Seeds

Materials Needed	water
seeds	lab equipment
container	dissection kit

1. ❑ Get a variety of seeds: maple, beans, lentils, corn and others.
2. ❑ Get a container to soak the seeds in.
3. ❑ Place seeds in container. Cover with water. Do not add a lid. Let soak overnight.

Next Day:

4. ❑ Get soaking seeds and lab equipment.
5. ❑ Slice a seed down the middle, long-ways.

When using a scalpel, always hold specimen with forceps - not fingers.

6. ❑ Can you identify the baby plant? ❑ Yes ❑ No
7. ❑ Can you see the leaves? ❑ Yes ❑ No
   How many leaves are there? ________________
8. ❑ Can you find the seed coat? *(The outer flexible layer of the seed, it usually falls off easily.)*
   ❑ Yes ❑ No
9. ❑ Using Lab Sheet 35 draw a picture of what you see with only your eyes.
10. ❑ Label.
11. ❑ Put the specimen under the microscope on low-power.
12. ❑ Draw a picture of what you see.
13. ❑ Label and fill in table.
14. ❑ Additional Lab: view the seeds through high-power. Draw, label and fill in table.
15. ❑ Clean up all equipment. Wash everything accordingly.

NAME ______________________

DATE ______________________

Plant Cells

Lab 21

Lab 21: Spores

Materials Needed
lab equipment
fern frond
moss

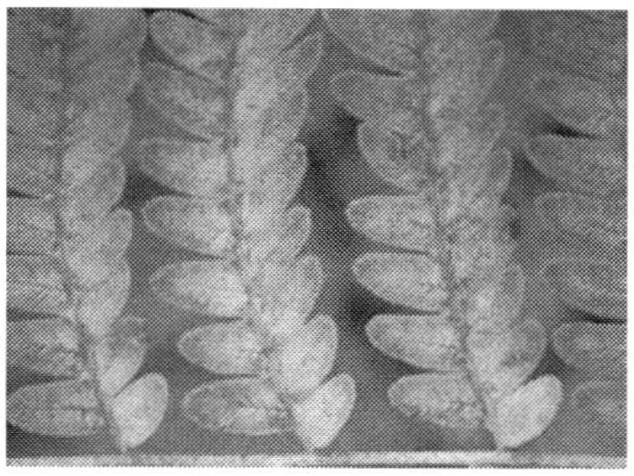

1. ❑ Get a fern frond.
2. ❑ Get out lab equipment.
3. ❑ Cut a small piece of frond and place it between a slide and a cover slip.
4. ❑ Focus it in the microscope on low-power.
5. ❑ Does it look like orange & brown grapes? ❑ Yes ❑ No
   Or does it look like thick, ropy coils of dark rings? ❑ Yes ❑ No
   (This will depend on the time of the year - it can look like either.)
6. ❑ Are there spores on the fern? ❑ Yes ❑ No
7. ❑ Using Lab Sheet 35 draw a picture of what you see. Label. Fill in the table.
8. ❑ Wash off slide and clean up equipment.
9. ❑ Additional Lab: view the frond under high-power. Draw, label and fill in table.

Next Day:

10. ❑ Get some moss.
11. ❑ Get supplies.
12. ❑ Cut a small piece of moss and place it between a slide and a cover slip.
13. ❑ Focus it in the microscope.
14. ❑ Using Lab Sheet 35 draw a picture of what you see. Label and fill in the table.
15. ❑ Wash off slide and clean up equipment.
16. ❑ Additional Lab: view the moss under high-power. Draw, label and fill in table.

NAME

DATE

Plant Cells

Lab 22

Lab 22: Mold

Materials Needed

1. ❑ Get a slice of lemon, lime, or orange.
2. ❑ Get a small bowl, a larger bowl and a box large enough to cover.
3. ❑ Place the slice of fruit on a small inverted bowl. See illustration A.
4. ❑ Place bowl in a larger bowl half full of water.
5. ❑ Put in an out-of-the-way place so it won't get knocked over.
6. ❑ Cover bowl with the box.

illustration A

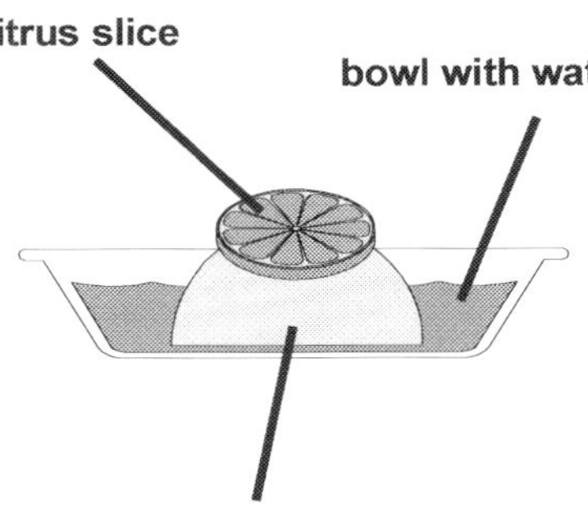

Three days later:

7. ❑ If fungus has grown thick and fuzzy, keep going with the experiment. If not, wait one or more days.
8. ❑ Get microscope, slides, alcohol, pipette, and mold-covered fruit.
9. ❑ Gently, scrape a bit of the fungus onto a slide.
10. ❑ Add a drop of water or alcohol with the pipette.
11. ❑ Focus the specimen in the microscope on low-power.
12. ❑ Does it look like groups of small mushrooms? ❑ Yes ❑ No
13. ❑ Does it look like a jumble of light-colored threads? ❑ Yes ❑ No
14. ❑ Can you see the dark spore cases (look like dots)? ❑ Yes ❑ No
15. ❑ Using Lab Sheet 35 draw a picture of what you see. Label and fill in table.
16. ❑ Focus the specimen in the microscope on high-power. (Remember to watch from the side.)
17. ❑ Do you see the spores? ❑ Yes ❑ No
18. ❑ Are they shaped like pollen? ❑ Yes ❑ No
19. ❑ Or are they shaped like seeds? ❑ Yes ❑ No
20. ❑ Draw a picture of what you see on high power. Label and fill in table
21. ❑ Clean up slide, put up equipment and bowls, throw away fungus and fruit.
22. ❑ Additional Lab: let the fungus on the fruit grow a few more days, then repeat experiment. Draw your observation, label and fill in table.

NAME ______________________________

DATE ______________________________

Plant Cells

Lab 23

Lab 23: Yeast

Materials Needed
lab equipment
yeast
sugar
small container
timer

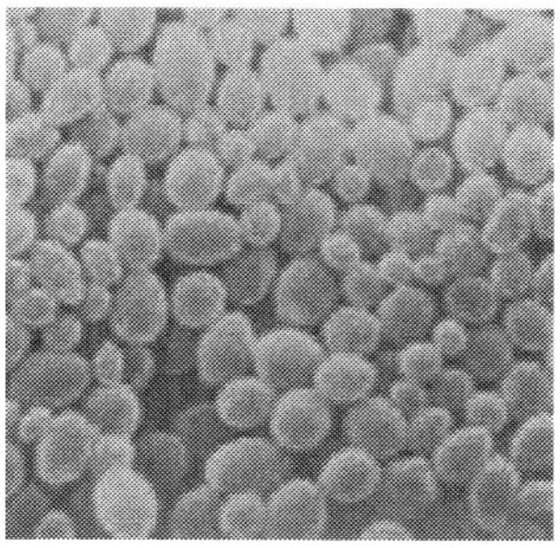

1. ❑ Get $^{1}/_{2}$ teaspoon of yeast and $^{1}/_{2}$ teaspoon of sugar.
2. ❑ Get a small container to mix in.
3. ❑ Get microscope, slides, covers, and timer.
4. ❑ Put 1 tablespoon of very warm water in the container.
5. ❑ Add $^{1}/_{2}$ teaspoon yeast and $^{1}/_{2}$ teaspoon sugar to water. Mix well. Wait 10 minutes (set timer).
6. ❑ Place a drop of mixture on a slide and carefully let down the cover slip on it.
7. ❑ Do you see a lot of air bubbles? ❑ Yes ❑ No
8. ❑ Do you see tiny shiny spheres around the bubbles? ❑ Yes ❑ No
9. ❑ Set timer for 2 minutes.
10. ❑ When 2 minutes is up, you should see movement, like a stream. Look at the outer rim of an air bubble.
11. ❑ Using Lab Sheet 35 draw what you see. Label and fill in table.
12. ❑ Looking at the microscope from the side, switch to a higher magnification.
13. ❑ Do you see the individual yeast cells? ❑ Yes ❑ No
14. ❑ If you see them, write a few words describing them: ______________________________
    ____
15. ❑ Focus on one spot to see buds forming on the parent cells. Do you see one?
    ❑ Yes ❑ No
16. ❑ Did the bud break off and float away on its own? ❑ Yes ❑ No
    This is how yeast cells reproduce.
17. ❑ Draw pictures of what you see on higher magnification. Label and fill in table.
18. ❑ Clean up the slides, container, teaspoon. Put up the equipment.

NAME

DATE

Plant Cells

Lab 24

Lab 24: Cork

Materials Needed
piece of cork
hand microtome or scalpel
lab equipment

1. ❑ Gather lab equipment and specimen.
2. ❑ Slice a small specimen of cork.
3. ❑ Place cork on slide.
4. ❑ Put slide on microscope and focus on low-power.
5. ❑ Use top lighting.
6. ❑ Can you see the individual cells?
7. ❑ What are the holes in the cork?
8. ❑ Below, draw a picture of what you see.
9. ❑ Label the picture.
10. ❑ Make your own table.
11. ❑ Repeat on high power. Draw observation. Label and write your own table.

Clean up all equipment. Wash everything accordingly.

NAME ....

DATE ______________________________________

Plant Cells

Lab 25

Lab 25: Wood

Materials Needed
lab equipment
pencil shavings

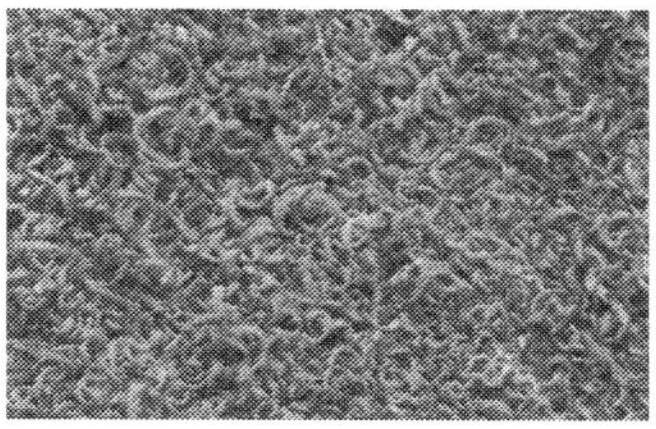

1. ❏ Gather lab equipment, including the pencil shavings.
2. ❏ View the pencil shavings under the microscope on low power.
3. ❏ Can you see closely-packed cells? ❏ Yes ❏ No
   What are these? ______________________________
4. ❏ Below, draw what you observe under the microscope.
5. ❏ Label what you can identify.
6. ❏ Can you see the xylem? ❏ Yes ❏ No
7. ❏ On the back of this sheet, briefly explain what xylem is.
8. ❏ Can you see the phloem? ❏ Yes ❏ No
9. ❏ On your own sheet of paper briefly explain what phloem is.
10. ❏ Additional Lab: do experiment on high-power. Draw observation. Label and write table.

____________________ ____________________

Clean up all equipment. Wash everything accordingly.

NAME ....

DATE ______________________________

Animal Cells

Lab 26

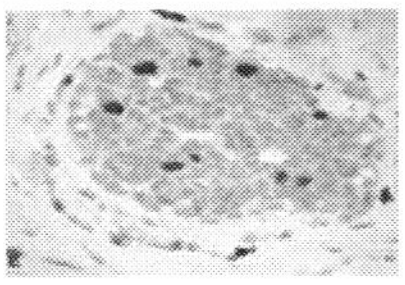

Lab 26: Epithelial Cells

1. ❑ **Materials Needed:** round-tipped toothpick, microscope, slide, cover slips, water, pipette and stain, iodine or food coloring.
2. ❑ Using the side of the toothpick (not the point), gently scrape the inside of your cheek.
3. ❑ Mount the scrapings on a clean, dry slide.
4. ❑ Add a drop of water.
5. ❑ Focus in the microscope on low power.
6. ❑ Do you see many clusters or sheets of irregularly shaped cells? ❑ Yes ❑ No
7. ❑ Focus on one particular cell. Can you see the nucleus? ❑ Yes ❑ No
8. ❑ Do you see the cytoplasm? ❑ Yes ❑ No
9. ❑ Do you see the cell membrane? ❑ Yes ❑ No
10. ❑ In the left circle below, draw a picture of what you see.
11. ❑ Remove the slide from the microscope. Place on paper towels.
12. ❑ Add a tiny drop of stain to the edge of the cover slip, letting it absorb under the cover slip.
13. ❑ Wait 2 minutes. Add 1-2 drops of water to rinse off the stain.
14. ❑ Can you see the nucleus better? ❑ Yes ❑ No
15. ❑ Can you see the cell membrane better? ❑ Yes ❑ No
16. ❑ When stained, bacteria look like small dark spots in the cell. Do you see any bacteria? ❑ Yes ❑ No
17. ❑ In the right circle, draw a picture of what you see after the stain.

Cheek Cells: No stain **Cheek Cells: Stained**

18. ❑ Wash off slide and clean up equipment.

NAME ______________________

DATE ______________________

Lab 27

Animal Cells

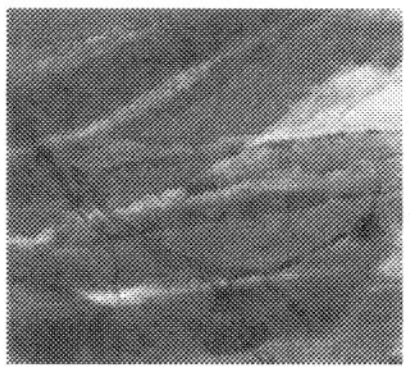

Lab 27: Muscle Cells

On Lab Sheet 35 draw samples of each specimen before and after stain. You may be able to obtain the animal parts from your local butcher. Call ahead about a week to order. It's usually free. As you work, draw two pictures of each specimen examined: plain and with stain. Label.

1. ❑ **Materials Needed:** slivers of uncooked (raw) steak, intestine, heart, cutting board, scalpel, forceps, microscope, slides, pipette and methylene blue stain (or blue food coloring) and iodine (or red food coloring).
2. ❑ Cut as thin a sliver of steak as possible. Use the forceps to spread it out.
3. ❑ Add a drop of stain.
4. ❑ Focus in the microscope.
5. ❑ Do you see very long bundles? ❑ Yes ❑ No
6. ❑ What type of cells are these? ______________________
7. ❑ Do you see the cell nucleus? ❑ Yes ❑ No
8. ❑ Do you see stripes in a band pattern going across the cells? ❑ Yes ❑ No
9. ❑ What type of cells are these? ______________________
10. ❑ Mount a piece of intestine or heart.
11. ❑ Add stain.
12. ❑ Focus the specimen in the microscope.
13. ❑ Are the cells larger or smaller than the steak cells? ❑ Larger ❑ Smaller
14. ❑ Now mount another piece and stain it red (iodine, food coloring or stain).
15. ❑ Can you see the single nucleus in each cell? ❑ Yes ❑ No *Look in the middle of the cell.*
16. ❑ Clean up, wash off, put up. Put specimen in the refridgerator. Draw pictures.

Next Day

17. ❑ **Materials Needed:** slivers of liver and kidney, cutting board, scalpel, tweezers, microscope, slides, and iodine or red stain.
18. ❑ Cut a thin sliver of liver and kidney. Use the tweezers to spread it out.
19. ❑ Focus on low-power. Draw what you observe. Label.
20. ❑ Add a drop of iodine or red stain.
21. ❑ Focus in the microscope. Think about how it is the same or different from yesterday's samples.
22. ❑ Using Lab Sheet 35 draw samples of each specimen, before and after stain.
23. ❑ Clean up, wash off, put up. Draw pictures.

NAME ______________________

DATE ______________________

Animal Cells

Lab 28

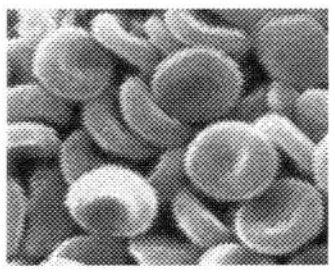

Lab 28: Blood Cells

Please take caution when obtaining blood, working with blood, and cleaning up the equipment.

1. ❑ **Materials Needed:** lancet, microscope, slides, timer, methylene blue stain (or blue food coloring, or iodine) and sterilized needle.
2. ❑ Using the needle, have an adult stick your finger to draw blood (or pick a scab).
3. ❑ Place a drop of blood near the end of a clean slide.
4. ❑ Take another slide and lower it until it touches the first slide at a 45° angle.
5. ❑ Spread the blood thin along the bottom slide with the top slide.
6. ❑ Focus in the microscope on low power.
7. ❑ Do you see red blood cells? ❑ Yes ❑ No
8. ❑ What is the proper word for red blood cells? ______________________
9. ❑ Can you see individual cells? ❑ Yes ❑ No
10. ❑ Draw a picture of a red blood cell. Do they have a nucleus? ❑ Yes ❑ No
11. ❑ Look at a picture to see a white blood cell. Can you see a white blood cell in the microscope? ❑ Yes ❑ No
12. ❑ What is the proper word for white blood cells? ______________________
13. ❑ Look in other areas to find one.
14. ❑ Take slide out of microscope. Add a drop of stain to the smear.
15. ❑ What do you see different about the white blood cells? ______________________

    ____
16. ❑ Do they have a nucleus? ❑ Yes ❑ No
17. ❑ Draw a picture of what you observed.
18. ❑ Clean up slide, equipment. Put up everything. 

Red blood cell ______________________

White blood cell ______________________

NAME ______________________
DATE ______________________

Fresh Water

Lab 29

Lab 29: Hunting for Microbes

Materials Needed
3 small jars with lids
(baby food jars work well)
water - 3 different sources
(without chlorine or chemicals)
mud, plants, grass or hay
labels

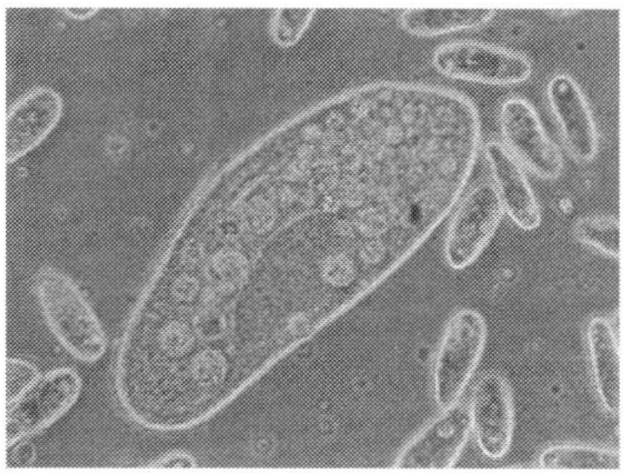

5 Days Later
microscope
well mount slide
cover slips
jars of water

At the end of this experiment, you may keep the water samples for a few more days to observe if more microbiotic life has grown.

1. ❑ Get out equipment.
2. ❑ Take jars outside to gather stagnant water. Get water (without chemicals or chlorine - use rain water or from puddles), mud & plants from 3 different places: pond, bucket, puddle, animal dish, rain gutter, or any other place that has standing water.
3. ❑ Put lids on jars.
4. ❑ Label each jar to remind you where each sample came from.
5. ❑ Take jars inside and put them in a dark place.

Five days later:

6. ❑ Get the jars of water (does it smell good?), microscope, well-mount slide, pipette, and covers.
7. ❑ Put a drop of specimen water on a well-mount slide.
8. ❑ Focus it under the microscope. Do you see any movement or signs of microbiotic life?
9. ❑ Using Lab Sheet 35 draw anything you see and try to label it. Look at reference books for identification.
10. ❑ Additional lab: let the water samples "ripen" a few more days, then repeat the experiment.
11. ❑ Are there more specimens? Draw and label.
12. ❑ Is there more variety? ❑ Yes ❑ No
13. ❑ Clean up, put up, toss water.

Clean up all equipment. Wash everything accordingly.

NAME

DATE

Sea Water

Lab 30

Lab 30: Diatoms

Materials Needed
lab equipment
small container of water
pipette
diatomaceous earth (DE)

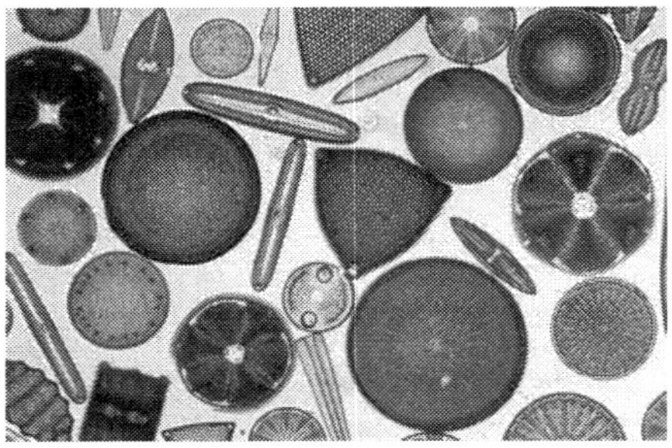

Diatomaceous earth (DE) is used in swimming pool water. If you have a pool, or a friend with a pool, ask for a teaspoon of DE.

1. ❑ Get microscope, slides, paper towels, pipette and a small container of water.
2. ❑ Get sample of diatomaceous earth (DE).
3. ❑ Put some diatomaceous earth on the slide.
4. ❑ Add a drop or two of water.
5. ❑ Put on a cover slip.
6. ❑ Focus it under the microscope on low-power.
7. ❑ Can you see anything?
8. ❑ Using Lab Sheet 35 draw a picture of what you see (you may draw more than one picture if you see more than one type or diatom).
9. ❑ Label the pictures. Fill in the table.
10. ❑ Additional lab: do this experiment under high-power. Draw observation. Label and fill in table.
11. ❑ Additional lab: make permanent mounts of diatoms.

Clean up all equipment. Wash everything accordingly.

NAME ____________________

DATE ____________________

Insects

Lab 31

Lab 31: Insect Legs

Make a copy of this page for each type insect used for the experiment.

Materials Needed

1. ❑ Get a large jar with a screw-on lid.
2. ❑ Put cotton ball soaked with ammonia *OR* cotton soaked with fingernail polish remover in the bottom. Put lid on. *(Do not mix chemicals.)*
3. ❑ Catch some insects: fly, grasshopper, bee, ant, beetle, butterfly, moth, mosquito.
4. ❑ Put insects in the jar with the cotton. They will die.
5. ❑ Get microscope, slides, scissors.
6. ❑ Use forceps to remove insects from jar.
7. ❑ Detach one leg from fly with scissors. Put it on a slide.
8. ❑ Focus the specimen in the microscope on low-power.
9. ❑ Is the leg hairy or smooth? ❑ hairy ❑ smooth
10. ❑ How many main parts does the leg have? ____________________
11. ❑ In a circle below, draw a picture of the specimen's leg. Label.
12. ❑ Look at the end of the foot. Do you see the small claws surrounding a soft cushion-like pad? ❑ Yes ❑ No
13. ❑ Do you see a lot of fine, sticky hairs on the pad? ❑ Yes ❑ No
14. ❑ How do the sticky hairs help the insect? ____________________
    ____
15. ❑ Do experiment with other insect legs. Draw what you observe, and label.
16. ❑ Clean up, put up. Put the rest of the insect bodies back in the jar for the next several experiments.

Specimen type: ____________________

Leg (with naked eye) | Foot | Leg (under microscope)

NAME ______________________

DATE ______________________

Insects

Lab 32

Lab 32: Insect Antennae

Make a copy of this page if there are more than six insects used for the experiment.

1. ❑ Get a mosquito and a moth if possible (look around window ledges, or near outside lights).
2. ❑ Put insects in the jar with cotton ball soaked with ammonia *or* cotton ball soaked with fingernail polish remover.
3. ❑ Get out the microscope, scissors, slides and deceased insect.
4. ❑ Dry mount and view the insect's head under the microscope.
5. ❑ Can you see the antennae segments? ❑ Yes ❑ No
6. ❑ Examine mosquito antennae. Is it a male or a female? ❑ Male ❑ Female
   Male will look like plumes (feathers), female will look like threads.
7. ❑ Look at a moth's antennae. Draw your observations and label.
8. ❑ Look at other insect antennae. Draw a picture of your observations. Label the picture to remind yourself which picture belongs to which insect.

______________ type of insect

______________ type of insect

______________ type of insect

______________ type of insect

______________ type of insect

______________ type of insect

NAME ______________________

DATE ______________________

Insects

Lab 33

Lab 33: Insect Wings

Materials Needed

1. ❑ Get out 5 insects, microscope, scissors, and slides.
2. ❑ Get out a butterfly. How many wings does it have? ______________
3. ❑ Make a dry mount of the butterfly wing.
4. ❑ Focus the wing in the microscope. Can you see the vein-like structure that makes the wings strong? ❑ Yes ❑ No
5. ❑ Does it have hairs or scales? ❑ hairs ❑ scales ❑ none
6. ❑ Do you notice anything else about it? Explain: ______________
7. ❑ Now look at a fly's wings. How many wings does it have? ______________
8. ❑ Focus the fly wing in the microscope. Can you see the veinlike structure that makes the wings strong? ❑ Yes ❑ No
9. ❑ Does it have hairs or scales? ❑ hairs ❑ scales ❑ none
10. ❑ Do you notice anything else about it? Explain: ______________
11. ❑ Look at the wings of different insects.
12. ❑ Fill in the chart on the next page with the following information about each insect:
    - ❑ the species of insect (stinkbug, cricket, honeybee, etc.)
    - ❑ number of wings
    - ❑ vein-like structures in wings, present or not (✓ if present)
    - ❑ on wings: hairs versus scales (write "hair" or "scales")
    - ❑ other (if you notice something else you want to include, do it!)
13. ❑ In the circles on the following page, draw a picture of the wing of each insect, as seen in the microscope. Label with species name.
14. ❑ Clean up, put up. Keep the insects for one more lab.

Insects: Wings
continued

Lab 33

Insect #1

Insect #2

Insect #3

Insect #4

Insect #5

Species	Insect #1	Insect #2	Insect #3	Insect #4	Insect #5
# Wings					
Veinlike Structures ✓ = yes					
Hair or Scales					
Other					

NAME ______________________

DATE ______________________

Insects

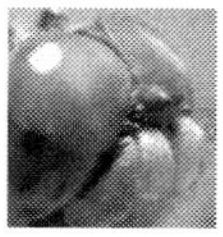

Lab 34

Lab 34: Insect Eyes

Materials Needed

1. ❑ Get out the insects, magnifying glass, microscope, probe, and slides.
2. ❑ Get out the fly. With a magnifying glass, look at its eyes.
3. ❑ Do the large oval eyes look like one eye or like many eyes? ❑ One ❑ Many
4. ❑ Do all the insect eyes look the same as the fly's eyes? ❑ Same ❑ Different
5. ❑ In the chart below, compare the number of eyes of each insect.
6. ❑ Give a brief description of the eyes: color, size, sheen, and texture.
7. ❑ Look at the butterfly. Look at its mouth. It is the curly part that is used to suck up food. It is called a proboscis. When it is not extended it lies coiled up in the form of a spiral under the butterfly's head. Did you see the mouth? ❑ Yes ❑ No
8. ❑ Look at a mosquito proboscis. Do you see the mouth? ❑ Yes ❑ No
9. ❑ On your own paper, compare and contrast the butterfly proboscis with that of the mosquito. With the probe, uncurl each and measure the length.
10. ❑ After you have finished the chart, clean up your area, throw away the insects, put everything in its place.

Species	Insect #1 ________	Insect #2 ________	Insect #3 ________	Insect #4 ________	Insect #5 ________
✓ One eye					
✓ Many eyes					
Color					
Size					
Sheen					
Texture					
Other					

Lab Sheet 35: Drawing Observations

Lab 35

Date		Specimen	Mount	Magnification	Measurement	Measuring Tool
	①					
	②					
	③					
	④					

① ______
label

Notes:

② ______
label

Notes:

③ ______
label

Notes:

④ ______
label

Notes:

Vocabulary

Define these words, or give their function, as they relate to a microscope:

Microscope Parts
100x
10x
40x
achromatic lens
adjustment lever
arm
base
body tube
chromatic aberration
clip
coarse adjustment knob
concave
convex
convex lens
convex mirror
cover slip
diaphragm
eyepiece
fine adjustment knob
high-power
lens
low-power
magnification
mirror
nosepiece
objective lens
ocular lens
ocular tube
oil immersion objective
opening
optic
revolving turret
slide
slip cover
stage
substage adjustment knob

Microscope Use
bottom lighting
depth of focus
direct sunlight
eyestrain
field of view
focal point
focus
focus up
grid
intensity
inverted image
invisible
light
light source
magnify
microscopist
nontransparent
opaque
reflected light
refraction
specimen
stage focusing
top lighting
transparent

Microscope Types
compound microscope
electron microscope
phase contrast microscope

Slides
dry mount
dye
Euparal
iodine
isopropyl alcohol
methylene blue
mount
mounting agents
neutral red
permanent slide
prepared slide
semipermanent slide
specimen
stage micrometer
stain
temporary mount
wet mount

Lab Equipment
crystallizing dish
forceps
graticule
microtome
petri dish
pipette
section
slide box
watch glass

Fibers
fiber
natural fiber
man-made fiber
scale
yarn

Crystals & Bubbles
boric acid
bubble
crystal
globule
three-dimensional
two-dimensional

Stains & Dyes
acid fuchsin
Carmine
color filter
dye
Eosin
iodine
isopropyl alcohol
methylene blue
microscopy
natural red
pulling stains
smear
stain
staining techniques

Cells
cartilage
cell
cell membrane
chromosome
cytoplasm
nucleus
animal cell
plant cell
chloroplast
granule
vacuole
cell wall

Plant Cells
anther
antibiotic
capillary action
chlorophyll
colony
cork
fern
filament
flower
frond
fungi
fungus
fruit
leaf
mildew
mold
moss
nectar
ovule
parent cell
petal
phloem
pistil
plant
pollen
reproduce
root
root hair
seed
spore
spore case
sporozoa
stalk
stamen
stem
vein
wood
woody stem
xylem
yeast

Animal Cells
cartilage
cell membrane
chromosome
cytoplasm
epithelial cell
"erythro"
erythrocyte
involuntary muscle
leukocyte
nucleus
red blood cell
smooth muscle tissue
striated muscle tissue
striated tissue
striations
voluntary muscle
white blood cell

Microbes
algae
ameba (amoeba)
animalcules
antibiotic
bacteria
biologist
botanist
chlorophyll
cilia
ciliates
colony (bacteria)
copepod
crustacean
desmid
diatom
diatomaceous earth (DE)
Euglena
flagellates
flagellum
food vacuole
food zone
hydroid
larvae
life cycle
marine
microbe
microorganisms
microscopic plant life
microscopic animal life
nucleus
organism
paramecium
parasite
polyp
protozoa
pseudopod
pseudopodia
rod
rotifer
sphere
spiral
sporozoa
stagnant
virus

Insect
antennae
compound eye
insect
mandible
proboscis
simple eye
spinneret
ammonia
locomotion
feeler
receptor
senses
covering
moth
butterfly
hair
scales
transparent

Scientific Suppliers

- Nasco, 901 Janesville Ave., Ft. Atkinson, WI 53538-0901 (800-558-9595)
- Science-Labs-In-A-Box, PO Box 1405, Canyon, TX 79015 (800-742-7805)
- Edmund Scientific, 101 E. Gloucester Pike, Barrington, NJ 08007-1380 (609-573-6250)
- Carolina Biological Supply Co., 2700 York Rd., Burlington, NC 27215 (800-334-5551)
- Tobin's Lab, PO Box 6503, Glendale, AZ 85312-6503

Answer Sheet

Lab 1: Microscope Parts

1. Eyepiece
2. Coarse adjustment knob
3. Fine adjustment knob
4. Arm
5. Substage adjustment knob
6. Base
7. Body tube or ocular tube
8. Objective Lens - 100X
9. Objective Lens - 40X
10. Objective Lens - 10X
11. Slide clip
12. Stage (with opening)
13. Diaphragm
14. Mirror
15. Nosepiece

Lab 2: Microscope Magnification

7. 100X **x** 10X = 1000X or
   1000 times magnification (larger than normal size)
8. 40X **x** 10X = 400X or 400 times magnification
9. 10X **x** 10X = 100X or 100 times magnification

Kym's Konstruction Kit

Kym's Konstruction Kit is fairly simple to use, just read the explanation to understand the system.

Below is a sample of the bottom of each lesson plan page. Write notes to yourself in the box, to help you remember. Every resource we need for a particular lesson, I write on this one page. It saves me having to remember where I wrote it down, where the picture is, what book it was in and all. So, if you write it all down on the bottom of the page, it will help you be organized!

A sample of the top of each lesson plan page is on the following page. It is a little more involved. The number in each boxed explanation correlates to the number below on the lesson plan sample. There is a line attaching the two. Beginning at ❶ read the explanations and look at the sample.

The lesson plans are on the following pages. If you would like to change a lesson plan, or set up your own - for one or more days - there is a blank page at the end for you to use. Make the schedule and lesson plans work for you and your family. Fit it into your schedule and needs.

Happy microscoping! *Kym Wright*

Bottom of lesson plan page:

Notes, videos, field trips, resources:

Daily
Read related material
Check for topical Websites
Answer questions
Copy Lab Sheets
Do Lab
Work on report

MU = Microscope Adventure Unit USB = *The World of the Microscope*

❺ **Daily** denotes the items that we do on a daily basis for this specific unit study. Our family begins the unit *after* we complete our core subjects (math, grammar, handwriting, and others) for the day.

"**Notes, videos, etc.**" is for you to write down videos to watch, materials to gather, resources you need, field trips for that topic, topical websites to visit or other notes to yourself. This is to help keep you organized with everything on one page.

This explains the abbreviations in the lesson plans.
MU = Microscope Adventure Unit, this curriculum.
You may add other abbreviations of resources used often. For example: if you are using *The World of the Microscope* by Usborne, you might list it on the right as "USB."

Kym's Konstruction Kit

Top of lesson plan page:

❶ The **date** you actually do the lesson. Example: 10/15 (Oct 15)

❷ A reasonable estimate of the **number of days** needed to cover this topic. Adjust accordingly.

❸ The **resource** used for this specific topic. Legend at bottom.

❹ **Page #**'s of the reference given. Example: Microscope Unit page 6 (MU 6).

❺ This shows the **topic** (noted in **bold**) of the day, so you can obtain topical resources and materials needed. Library listings are on page 9.

❶ Date	❷ # Days / Topic	❸ Book	❹ Pages	❺ Topic / Lesson Plans	❻ Copy	❼ Vocabulary	❽ Other Resources	❾ Page #
	1			**Setting Up**				
		MU	6	Set up Microscope Notebook				
		MU	7-8	Gather lab equipment				
		MU	7-8	Gather lab materials				
	2			**Microscope Parts**		✓		
		MU	10	Read about microscope parts				
		MU	10	Answer questions				
		MU	38	Do Lab 1	38			
		MU	10-11	Read about the stage & lighting				
		MU	11	Read about objectives & magnification				
		MU	10-11	Answer questions				
		MU	39	Do Lab 2	39			

❾ The **page #** of the "Other Resources."

❽ **Other resources** you need or have: library books, lab materials, etc.

❻ * If you need to **copy** a Lab Sheet or other workpage, list it here.

❼ A reminder for you to daily check and work on the **Vocabulary** words for the lesson. ✓ is to remind you to check it off daily. **Vocabulary words** are on pages 75-76.

* Remember to look on page 8 for the "Lab Sheets to Copy" list.

Date	# Days / Topic	Book	Pages	Topic / Lesson Plans	Copy	Vocabulary	Other Resources	Page #
	1			**Setting Up**				
		MU	6	Set up Microscope Notebook				
		MU	7-8	Gather lab equipment				
		MU	7-8	Gather lab materials				
	2			**Microscope Parts**		✓		
		MU	10	Read about microscope parts				
		MU	10	Answer questions				
		MU	38	Do Lab 1	38			
		MU	10-11	Read about the stage & lighting				
		MU	11	Read about objectives & magnification				
		MU	10-11	Answer questions				
		MU	39	Do Lab 2	39			
	1			**History of Microscopes**				
		MU	12	Read about the history				
		MU	12	Answer questions				
		MU	5, 75-76	Choose research topic or person (use vocabulary words for more research ideas)				
				Begin research				

Notes, videos, field trips, topical websites, resources:

Daily
Read related material
Check for topical Websites
Answer questions
Copy Lab Sheets
Do Lab
Work on report

MU = Microscope Unit

Date	# Days / Topic	Book	Pages	Topic / Lesson Plans	Copy	Vocabulary	Other Resources	Page #
	1			**Set Up, Use & Clean Up**		✓		
		MU	13	Read about setting up the microscope and lab equipment				
		MU	40-41	Do Lab Sheet 3	40-41			
			13-14	Read about using the microscope, low power & high power				
				Read about drawing observations				
		MU	74	Do Lab Sheet 35, with prepared slide	74			
		MU	14	Read about cleaning up labs				
		MU	41	Clean up lab				

Notes, videos, field trips, topical websites, resources:

Daily
Read related material
Check for topical Websites
Answer questions
Copy Lab Sheets
Do Lab
Work on report

MU = Microscope Unit

Date	# Days / Topic	Book	Pages	Topic / Lesson Plans	Copy	Vocabulary	Other Resources	Page #
	4			**Lab Equipment & Technique**		✓		
		MU	15	Read about types of slides				
		MU	15	Answer questions				
		MU	15	Read about slide mounts				
		MU	15	Answer questions				
		MU	15	Read about wet mounts				
		MU	42	Do Lab Sheet 4	42			
		MU	15	Read about dry mounts				
		MU	43	Do Lab Sheet 5	43			
		MU	15	Read about permanent and semi-permanent mounts				
		MU	44	Do Lab Sheet 6	44			
		MU	15	Read about well mounts				
		MU	45	Do Lab Sheet 7	45			
		MU	16	Read about other lab equipment				
		MU	16	Answer questions				
		MU	16	Read about microtomes				
		MU	16	Answer questions				
				Make a hand microtome				

Notes, videos, field trips, topical websites, resources:

Daily
Read related material
Check for topical Websites
Answer questions
Copy Lab Sheets
Do Lab
Work on report

MU = Microscope Unit

Date	# Days / Topic	Book	Pages	Topic / Lesson Plans	Copy	Vocabulary	Other Resources	Page #
	3-4			**Fibers**		✓		
		MU	**18**	Read about fibers, fabric and yarn				
		MU	**18**	Answer questions				
		MU	**46**	Do Lab Sheet 8	46, 74			
		MU	**18**	Read about paper				
		MU	**47**	Do Lab Sheet 9	47, 74			
		MU	**18**	Read about hair, fur, feathers and skin coverings				
		MU	**48**	Do Lab 10	48, 74			
	2			**Film & Dust**				
		MU	**19**	Read about film and top lighting				
		MU	**19**	Answer questions				
		MU	**49**	Do Lab Sheet 11	49, 74			
		MU	**19**	Read about dust				
		MU	**50**	Do Lab Sheet 12	50, 74			

Notes, videos, field trips, topical websites, resources:

Daily
Read related material
Check for topical Websites
Answer questions
Copy Lab Sheets
Do Lab
Work on report

MU = Microscope Unit

Date	# Days / Topic	Book	Pages	Topic / Lesson Plans	Copy	Vocabulary	Other Resources	Page #
	2-3			**Crystals & Bubbles**		✓		
		MU	**20**	Read about crystals				
		MU	**20**	Answer the questions				
		MU	**51**	Do Lab Sheet 13	51, 74			
		MU	**20**	Read about bubbles				
		MU	**20**	Answer questions				
		MU	**52**	Do Lab Sheet 14	52, 74			
	2			**Stains & Dyes**				
		MU	**21-22**	Read about stains & dyes				
		MU	**21-22**	Answer questions				
				Practice stains	74			

Notes, videos, field trips, topical websites, resources:

Daily
Read related material
Check for topical Websites
Answer questions
Copy Lab Sheets
Do Lab
Work on report

MU = Microscope Unit

Date	# Days / Topic	Book	Pages	Topic / Lesson Plans	Copy	Vocabulary	Other Resources	Page #
	6-7			**Cells & Plant Cells**		✓		
		MU	23	Read about cells				
		MU	23	Answer questions				
		MU	53	Do Lab Sheet 15	53			
		MU	23	Read about plant cells				
		MU	23	List major plant parts				
		MU	54	Do Lab Sheet 16	54			
		MU	23	Read about flowers				
		MU	23-24	Answer questions				
		MU	55	Do Lab Sheet 17	55, 74			
		MU	24	Read about plant water ducts				
		MU	24	Answer questions				
		MU	56	Do Lab Sheet 18	56, 74			
		MU	24	Read about pollen				
		MU	24	Answer questions				
		MU	57	Do Lab Sheet 19	57			
		MU	25	Read about seeds & fruit				
		MU	25	Answer questions				
		MU	58	Do Lab Sheet 20	58,74			

Notes, videos, field trips, topical websites, resources:

Daily
Read related material
Check for topical Websites
Answer questions
Copy Lab Sheets
Do Lab
Work on report

MU = Microscope Unit

Date	# Days / Topic	Book	Pages	Topic / Lesson Plans	Copy	Vocabulary	Other Resources	Page #
	6-7			**Other Plants**		✓		
		MU	**26**	Read about ferns & spores				
		MU	**26**	Answer questions				
		MU	**59**	Do Lab Sheet 21	59, 74			
		MU	**26**	Read about fungus, mold & mildew				
		MU	**26**	Answer questions				
		MU	**60**	Do Lab Sheet 22	60, 74			
		MU	**27**	Read about yeast				
		MU	**27**	Answer questions				
		MU	**61**	Do Lab Sheet 23	61, 74			
		MU	**27**	Read about Cork				
		MU	**27**	Answer questions				
		MU	**62**	Do Lab Sheet 24				
		MU	**27**	Read about wood				
		MU	**63**	Do Lab Sheet 25				

Notes, videos, field trips, topical websites, resources:

Daily
Read related material
Check for topical Websites
Answer questions
Copy Lab Sheets
Do Lab
Work on report

MU = Microscope Unit

Date	# Days / Topic	Book	Pages	Topic / Lesson Plans	Copy	Vocabulary	Other Resources	Page #
	6-7			**Animal Cells**		✓		
		MU	28	Read about animal cells				
		MU	28	Answer questions				
		MU	64	Do Lab Sheet 26	64			
		MU	28	Read about muscle cells				
		MU	28-29	Answer questions				
		MU	65	Do Lab Sheet 27	65			
		MU	29	Read about blood cells				
		MU	29-30	Answer questions				
		MU	66	Do Lab Sheet 28	66, 74			
				Microbes				
		MU	31	Read about fresh water microorganisms				
		MU	31	Answer questions				
		MU	31	Draw pictures of common microorganisms; label				
		MU	67	Do Lab Sheet 29	67, 74			
		MU	32	Read about microscopic marine life				
		MU	32	Answer questions				
		MU	68	Do Lab Sheet 30	68, 74			
		MU	32	Read about microscopic life in soil				

Notes, videos, field trips, topical websites, resources:

Daily
Read related material
Check for topical Websites
Answer questions
Copy Lab Sheets
Do Lab
Work on report

MU = Microscope Unit

Date	# Days / Topic	Book	Pages	Topic / Lesson Plans	Copy	Vocabulary	Other Resources	Page #
	6-7			**Insects**		✓		
		MU	**33**	Read about insects				
		MU	**33**	Answer questions				
		MU	**33**	Read about insect legs				
		MU	**33**	Answer questions				
		MU	**69**	Gather insects & do Lab Sheet 31	69			
		MU	**33**	Read about insect antennae				
		MU	**34**	Answer questions				
		MU	**70**	Do Lab Sheet 32	70			
		MU	**34**	Read about insect wings				
		MU	**34**	Answer questions				
		MU	**71-72**	Do Lab Sheet 33	71-72			
		MU	**35**	Read about insect eyes				
		MU	**35**	Answer questions				
		MU	**73**	Do Lab Sheet 34	73			
		MU	**35**	Read about insect mouths				
		MU	**35**	Answer questions				
		MU	**35**	Draw a butterfly mouth				
	1	**MU**	36	**Ending the Unit**				

Notes, videos, field trips, topical websites, resources:

Daily
Read related material
Check for topical Websites
Answer questions
Copy Lab Sheets
Do Lab
Work on report

MU = Microscope Unit

Date	# Days / Topic	Book	Pages	Topic / Lesson Plans	Copy	Vocabulary	Other Resources	Page #

Notes, videos, field trips, resources:

MU = Microscope Unit

Neat Sites on the Web

Websites and On-line Resources

The on-line world is tentative and transitory. What is here today, could very well be gone tomorrow. These listings were available in the summer of 2007.

A word of caution about using the on-line world. When researching any topic, unacceptable websites might come up in the listings. If students will be using the internet, you can make the filtering system on your browser more restrictive. Just please use caution. And don't let children search alone! The websites listed here were looked through, and we found nothing questionable about them. *We cannot guarantee them, nor any links off of these websites*. Our suggestions are for an adult to view the websites first, and to stick with the well-known names and sites.

There is a lot of information, hints, tidbits, suggestions, examples and more on these web-sites. They are chock-full of helps.

Enjoy riding the surf!

Microscope Lab Review http://www.ou.edu/research/electron/mirror/index.shtml

Microscope Parts Tests http://www.stanford.edu/group/Urchin/-pom.htm

Citric Acid Crystals (viewed through a microscope) http://www.tssphoto.com/abstract/EABS0412.html

Robert Hooke http://www.comlab.ox.ac.uk/archive/ox/univ-col/boyle-hooke.html

History of the Microscope http://www.southwestschools.org/jsfaculty/Microscopes/history.html

Micoscopy in the United Kingdom - http://www.microscopy-uk.org.uk/

Drawing through the Microscope http://shs.westport.k12.ct.us/mjvl/biology/microscope/microscope.htm

Art in Science - crystals - http://www.lci.kent.edu/polmicpic.html

Foods Under the Microscope http://www.magma.ca/~scimat/

Minerals Under the Microscope http://www.mineralarts.com/micromin/micromin.html

Microscapes: Images From the Microscope - http://micro.magnet.fsu.edu/

Insects Under the Microscope http://www.pbrc.hawaii.edu/bemf/microangela/

Microscope Slides http://www.lehmanscientific.com/p_slides.html

Micro notes

Micro notes

"The Microscope Adventure! will be a very useful one to homeschoolers or to regular schools.

Kym Wright has done a great job on it. The lab write-ups will make it easy for instructors to copy for the family as required and then slip the paperwork into the biology notebook. Or students could just use it as it is. The really good part for junior high schoolers is that the labs use materials which usually come with the scope or can be easily gotten, no expensive prepared slides here. Many of these labs can also be used for high school homeschoolers. (I shouldn't say that since I wrote a biology lab manual for high school level, right?) Oh well, this unit on microscopy is really good and a bargain. I definitely recommend it."

Kathleen Julicher - Biologist, homeschooling mom, columnist for *Homeschooling Today,* Author of 3 high school lab texts: *"Experiences in ..."* Chemistry, Biology & Physics ~~ by Castle Heights Press.

Microscope Adventure!
by Kym Wright

Learn & Do Unit Studies

by alWright! Publishing
P.O. Box 81124
Conyers, Georgia 30013

Kathleen Julicher is a biologist, homeschool mom of four (3 are in college), columnist for *Homeschooling Today Magazine*, and author of the high school science lab series *"Experiences in ..."* Chemistry, Biology and Physics.

1. First printed in *Homeschooling Today* Nov/Dec 1993, where you can read a review of various microscopes.Used by permission from: Castle Heights Press, Inc., 2578 Alexander Farms Drive SW, Marietta, GA 33064. E-mail: Julicher@aol.com. Website:http://www.flash.net/~wx3o/chp. Read another microscope review in *Homeschooling Today* Jan/Feb 1997.

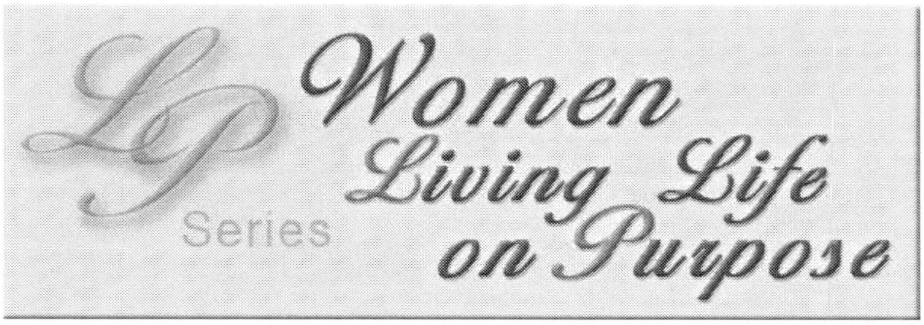

In the ***Women: Living Life on Purpose Series***, we have CD sets and companion books on various topics.

In her talks, Kym casts a vision of motherhood, being a wife, organizing the house, and loving it all. She shares scripture, practical solutions, and all the warm stories she's known for. Each topic is well researched with biblical concordances, commentaries and the Bible itself, along with information from the experts.

Come join Kym as she shares her heart for women to live life on purpose.

Children & Chores

3-CD Set & Companion Book

From a "let's enjoy ourselves and have fun" and the work not getting done, to a more organized family life, join Kym as she shares her journey in teaching her eight children how to work and choose a good attitude about it.

She talks about children's chores from a biblical perspective, casting a vision of why we want to teach our children to work. What the Bible has to say about work and training our children. Kym helps you establish the reasons to teach our children to work.

Work attitudes are discussed – ours and our offspring's – and how to create the plan with chore charts.

Family Meals

3-CD Set & Companion Book

Research shows when adults and children eat together . . .children do better in school, they have fewer behavior problems, and they like talking with the adults in their lives. Teens are less apt to smoke, use alcohol or drugs and family values and traditions and traditions are passed on.

Studies indicate that meals are more nutritious and healthful, food dollars stretch further, time is better utilized because the family only cooks one meal, and many times the family is involved in meal preparation. Are we talking about "The Waltons," or will this work for me and my family? Join Kym as she shares the vision of the **Family Meal**, its importance, how to plan meals and shop for groceries with great ease, and how each family can create special times around the table, with just a bit of preparation.

Moms Mentoring Moms

This series came about through Bible Studies Kym held in her home, for young mothers and single women. When they finished our study on Children and Chores, and on Family Meals they asked Kym to teach them a short study on Titus 2. Many call these verses the "Older women teach the younger women verses." Kym just calls them the "Moms Mentoring Moms" scriptures.

What do these verses mean? How do they apply to my life today? Can we really find practical guidance in these scriptures? We'll answer these questions and more.

Kym shares the Don't's, the Do's and the Why of Titus 2: How are we to live? If I don't drink, then how can that portion of the verse apply to me which says not to be enslaved to much wine? With a strongly biblical basis, and many of her heart-warming stories, Kym shares the very practical ways we can live out these verses in everyday life. Without being Superwoman, and with the husband and children we each have.

Come and join Kym as she shares about Moms Mentoring Moms.

Creating Creativity

1-CD Set

"An idea can turn to dust or magic, depending on the talent that rubs against it."
~ Bill Bernbach, advertising expert

Why do we, as moms, want to invest our time and money in art projects? Especially when we already have so much on our plates. Why would we want to even delve into this expanse of art? Where the lines are blurred, and we're not even experts?

Many moms feel that they have the proverbial "2 left feet" when it comes to art. To set a foundation for learning, Kym references some of the greatest creative forces in our world: Einstein, Bill Gates, the founder of Poloroid Corporation, artists, educators, philsophers, and doctors. All agree that creativity is a vital element in life, and teaching it to our children is integral to their future success.

Join Kym as she shares ways to teach our children creativity – even if that trait is not ours. She gives ideas, resources, lists and more to help you appear more creative, and to help our children accomplish it.

Gift of Goals

1-CD Set

Research indicates those who write their goals out, 95% of them will reach their goals. The Bible gives reason to plan our lives in several places: Proverbs 29:18 "Where there is no vision, the people perish." And Isaiah 32:8, "But the noble man devises noble plans; and by noble plans he stands."

Planning keeps us from acting in haste – from being so spontaneous that it gets us in trouble. Or that we flit from this to that, and never settle in on any one thing. And how do we share this with our children? How do we help them live by goals, rather than just responding and reacting to life?

Join Kym as she shares her journey of the Gift of Goals: the areas to set them in, how to reassess and how often, and how to help our children dream big, and break these into smaller, bite-size pieces.

Magazine Article Writing

6-CD Set & Session Notes

Let the story inside you become a blessing to others

A course on getting published

Turn your dreams into bylines and help yourself to a bright future as a magazine writer. If you're a determined writer – experienced or not, this class will provide you with the skills you need to get published.

This course is about how to write for magazines, getting published and to broaden your scope while helping yourself to a bright future as a magazine writer. Learn to share your stories and testimony, God's blessing in your life, and encouragement to others. You will learn how to find interesting ideas and angles for articles to locate suitable markets, how to research and write query letters, article structure and submitting articles – all this and more.

This series will jump start your career. You'll learn plenty of powerful brainstorming techniques designed to practically write every article for you: where to get ideas, how to structure your articles, how to find markets, and how to make queries and submissions. We'll take a closer look at the magazine editorial process and how you can make that work to your advantage by focusing your writing for a particular audience. You'll learn how to approach major market magazines, how to get past the slush pile, how to follow up until you have a sale and how to follow through until you have completed the assignment satisfactorily. Then we'll discuss making the most use of your time: time management, how to resell your article, and how to write a second article from the same research.

Kym Wright

Kym is an energetic and encouraging speaker. She has spoken to many groups including MOPS or Mothers of PreSchoolers groups, state homeschool conventions, local homeschool groups, Rotary and Kiwanis Clubs, Civic Organizations, church groups, women's events and seminars, teen groups, and more!

Kym is the mother of eight homeschooled children. She and her husband, Mark, have homeschooled since the mid 80s. They have graduated four students who are attending college and holding down jobs, internships and apprenticeships in their chosen fields.

Sharing her love of motherhood and homeschooling on **The 700 Club**, on **radio** shows and in her writing, **her passion shows . . . and it's contagious**.

Kym is author of the **Learn & Do Unit Studies** and ***Women: Living Life on Purpose***, a mommy-primer helping women become the best wife, mother and friend they can be. She has been the editor of various local, regional and statewide newsletters, as well as having her articles published in many homeschool publications, including *Homeschooling Today, The Old Schoolhouse,* and *Practical Homeschooling* and now publishes an online magazine ***The Mother's Heart***.

Kym writes to encourage moms in very practical, hands-on ways making the philosophical very workable. She shares the love of raising children, teaching preschoolers, being the mother of older children, homeschooling and the privilege of helping our children find God's call in their lives.

Visit her websites:

www.KymWright.com
www.Learn-and-Do.com
www.The-Mothers-Heart.com
www.Living-Life-on-Purpose.com

To contact Kym to speak for your event: info@KymWright.com or Kym Wright, PO Box 81124, Conyers, Ga 30013.

The Mother's Heart Back Issues

#1Aug 96

Vol 1 #1: Aug '96: Letters From Your Heart; Coming Home (Kym's journey from business to staying home); Parenting Power: The Foundation; Thinking about Adoption; Passing On Harvard; Life on Schedule; Becoming the Family Herbalist; Keeping up with the Jones'; Miracle of Morgan; When Daddy Travels.

#2 Nov 96

Vol 1 #2: Nov '96: Life Influencers; Respect; What Pushes Your Buttons?; Adoption: HomeStudy and Kym's Experience; Infertility... and God; Literature: Literacy or Legacy?; Rising Early; Natural Pregnancy; Home HairCutting; A Medical Couple Looks at God's Family Planning; Families ... like Quilts; The Hospitable Home; I Don't Know How You Do It!; Cows; Milk; Cheesemaking.

#3 Feb 97

Vol 1 #3: Feb '97: A Joyful Mother; Obedience; Who's In Charge Anyway?; Adoption: Consents; Breast Infections; Straight Talk; Finding Extraordinary in Ordinary; Male's HairCut; Children's Chores; Choosing Contentment; Herbs: Mommy Diagnostics; Ready for a Baby; Million Dollar Family; Loneliness; SuperMom!

#4 May 97

Vol 1 #4: May '97: Country Chic!; Her Children Rise Up and Call Her Blessed; From Your Heart; Responsibility; First Things First!; Birth Father's Rights; Why Adopt?; Pregnant With #10!; Learning To Trust: Mom of 11; Mothering on Purpose; Children's Chores (part 2); Blank Chore Chart; Internet & Heart; Cheap Dec! & Painting Verdé; Herbs: PMS: Pre-Menstral Syndrome: What Is It?

#5 Aug 97

Vol 2 #5: Aug '97: Bits 'N Pieces; Behind Every Successful Mother; Discerning Vital Life Principles: Mate Maintenance; Biracial Adoption; His Way; Infertility: Desire of My Heart; Private Decisions; Color Blind; Autumn Garden; Our Family of 14 Children; How Can We Trust God More For Family Planning?; Lavish Living; Kitchen ShortKuts; Our New Kitchen; Female's Blunt HairCut; Garlic Bag; Pattern Notebooks; Older Siblings; Staying Home While Sick; Library: @ Your Service, Onsite & Form; Bird Unit Study.

#6 Nov 97

Vol 2 #6: Nov '97: A Higher Education; Family Life; Spanking; Wiring 101; Discussing Adoption Concerns; Gentle Schedules; Blending Natural & Modern Medicine; His Queen; My King; Infertility & God; VP to Play Dough Mom; Scrunchies; Alter a Bathrobe; Children's Measurements & Forms.

#7 Feb 98

Vol 2 #7: Feb '98: Dealing with Disappointment; Family Fun; Capturing Their Heart; Wiring 102; Open & Closed Adoption; Dominoes!; Mother's Peace; From Australia; Finding Time; Baking Bread; Orange Raisin Bread; Meal Preparation; Chicken Dishes; Incubating Eggs; Shutting Down the Complaint Dept.

#8 May 98

Vol 2 #8: May '98: Faith or Denial?; A Home Much-Loved; Capturing Their Heart; Charting a Life Course; Adoption '98; Double Blessing; More Children ~ Adoption?; Herbs; Eating Better: The Course; Preschool Basket; School Time-Management & Form; Journey to

Contentment; Decorating Notebook; If We Listen.

#9 Aug 98

Vol 3 #9: Aug '98: A Matter of the Heart; The Harvest; Charting A Life Course; When Baby Dies in Your Womb; Recipe Experimentation; Eating Better ; Beef Dishes; Tips & Tricks; Creating Creativity; More Fun Stuff Ideas!; Once Upon A Time: Child Abuse; A Love for our Children; Trust; "Spoil"; Why are Good Books So Important; The Rest Can Wait; Women: Living Life on Purpose

#10 Nov 98

Vol 3 #10: Nov '98: Children's Life Goals; The Power of the Spirit; Charting A Life Course, Pt 3; Garden of Health; Making Good Better; Faith; Growing In Grace; In Sickness or In Health; Preschool Literature; Schooling Many; One Man's Twaddle; School Organization; The Wall; Committed or Obligated? Botany; Joys of Staying Home; Ambassador of Motherhood

#11 Feb 99

Vol 3 #11: Feb '99: Moving & Change; Mighty in Spirit; Adolescence; From White to Whole; Longing for an Open Womb; Adoption; Homeschooling & Organization; 'Net Connected; Correspondence College; Blessing & Dressing; Love Letter; A Journal Entry; Counting the Cost; Home & Ministry.

#12 May 99

Vol 3 #12: May '99: Coming Home ... Again; Quiet Life; Joyful In Spirit; Children are a Blessing; Eating Better: Y2K & Beyond; Hyperchilderemia; Interstate Adoption; Raising Biological & Adopted Kids; Gentle Beauty; Preparation & Organization; Homeschooling Special Needs Children; Dean to Mom; The Professional Woman I Married; Stepping Up Your Child's Space; Scripture Memorization; A Word Kindly Spoken

#13 Aug 99

Vol 4 #13: Aug '99: Standing Watch; Progress; Adolescence; Four Food Storage Plans; Menopause; Family Fitness Fun; Nausea Remedies; Healthy Granola; International Adoption; Adoption & Others; The Wait; Growing in Grace; Is the Husband Enough?; The Waltons Revisited; Classified Ads; Press On; Letting Go; A Woman of Virture; G.E.R.M. Warfare!

#14 Nov 99

Vol 4 #14: Nov '99: Maintaining the Cover; "I Will"; The Father's Blessing; Is the Husband Enough? Pt 2; Grain Mills; Immuno-Woes; Teenage Birth-mothers; Wrong Reasons to Adopt; Babies Come in Many Ways; Quilting 101; Bringing Your Heart Home; Smiley Face; Older Mom; She Rises Before Dawn... to pray; Submitting; Creation Science; Good Books; Photography Unit

#15 Feb 00

Vol 4 #15: Feb '00: Equipping for Excellence, Working Women, Dating/ Courtship Parameters, Settling Family Convictions, Pastry Grains, Having Healthier Children, Regaining Pre-Pregnancy Figure, Our Story of Grace (Adoption), Time Management, Homemaker or Housewife?, Learning Takes Humility, Scrapbooking, Good Books, Making Units (Studies) Yours; Scripture Memorization... Psalm 91, New Unit: Flower Arranging & Wreaths

#16 May 00

Vol 4 #16: May '00: Discernment; Dailyness; The *Family Pow-Wow*; How Do I Get My Husband to Lead?; Grain Variety for Quick Bread; Too Tired to be Exhausted?; Getting Started in Adoption; Honoring My Husband; Finally ... Children; Moving; Titus 2:4-5 — What Does It Really Mean?; Building up Child's Dreams; Homeschooling with Preschoolers; Why We Homeschool

#17 Aug 00

Vol 5 #17: Aug '00: Music & Our Children; Brown Rice; Defeating Depression Naturally; Adoption Protection; Foster Parenting; Vaccinate or Not?; Stretch Travel Dollar; Unequally Yoked; LifeGuard Mom; Love Letter; Chemistry; Dissection, Autism, PDD & Celaic Disease; Recipes; Unit Study Answers from Authors

#18 Nov 00

Vol 5 #18: Nov '00: Shepherding; Count Your Blessings; Basketball, Football & Music; Coaching Tight; Thanksgiving Favorites; Defeating Depression, Pt 2; Candida Yeast; Trusting the Creator; Foster Parenting; Shepherd's Crook;

Adoption Story; Children ... a Gift; Spiritual Housecleaning; Wright Photo Gallery

#19 Feb 01

Vol 5 #19: Feb '01: Serving Others; Unsaved Husbands; Whole Grain Crepes; ADD / ADHD; Bed-Wetting; Candida Questions; Weigh Down Update; Loving Baby Doe; Foster Care Resources; Tightwad Resolutions; Rising & Time with Husband: Q&A; Dream House; Special Children & Special Moms; Tag-A-Long Meals; Recipes; Baby Showers

#20 May 01

Vol 5 #20: May '01: Good-bye; An Engagement Story; Friends! Really?; Pasta Salad Italian; Cancer Prevention Tips; Cook Books; **;** Keeping Our Arms Open; Becoming Your Child's Advocate; Adopting Your Foster Child; Sincerely Wondering; Frugal & Organized; Magazines; House & Garden; IEP: easy as PIE

#21 Feb 05

Vol 6 #21: Feb 05: Some Answers: A Love Letter; E-World; Catching up with the Wrights; Whole Grains: 2-Stage Process; Whole Wheat Bread; Moms Mentoring Moms; Grace in Teen Relations; Values; Empowering the Pastor's Wife; Adoption: My Spiritual Journey; Teaching Art at Home; Making Unit Studies Yours; Creation Science; Science Activities for Children; America's Junior Miss; Keepers at Home

#22 April 05

Vol 6 #22: April 05: Husband, Father, Protector: Destroyer of Lies; Purring & Italy; Dairy Dilemma; Bread-U-Cation; Awakening Love Too Soon; The Joy of Motherhood; The Expert Enabler; Finding Agreement in Adoption; Why Teach Art: Making Unit Studies Yours Part 2; Informal Learning; Coming Out of My Shell

#23 June 05

Vol 6 #23: June 05: Pulling Close; Do This; Whole Grain Fiber; The Social Needs Child; Lessons from Samson & Delilah; Behind the Scenes Ministry; Bartering; Adoption: Living with "No"; Purchasing an Art Curriculum; Making Units Yours Part 3; Rewards of Volunteering

Vol 6 #24: Aug 05: Boys will be Boys; President of the Social & Fitness Club; My Son is Learning to Cook!; Unit Yak; Online Forum; The Important; Coconut Oil; Choosing Healthy Oils; Words & Attitudes Matter; Forgiveness; Proverbs 31: The Virtuous Woman; Trust; Adoption: The Home Study; 1 Corinthians 13 for Busy Moms; Art Supplies & Materials; Making Unit Studies Yours Part 4; A Year of Cooking

#24 Aug 05

#25 Oct 05

Vol 6 #25: Oct 05: Enemy at the Gate: Overseeing your Wife in Perilous Times; Raising Boys & Girls to be Grown Men & Women; Eating on the Run; Proverbs 31: Virtue; The Male Hormone Question; At-Home Mom; Adoption: The Role of Friends; Art: Age Level Criteria; All About Lap Books; Making Units Yours Part 5; Books on Tape; Unconventional College Credit; Meeting the Needs of Homeschoolers; Unit Yak; Weekly Wakeup with Kym Wright

#26 Dec 05

Vol 6 #26: Dec 05: The Worth of a Mom at Home; Christmas Traditions & Holiday Baking; Sweet Holiday Memories; Proverbs 31: An Excellent Wife; Christmas Preparation; Sharing our Failures; To The Wonderful Friends in My Circle; Seasons; Private vs Agency Adoption; Her Heart; Art: Studying from Nature; Unit Yak: Family Unity in Studies; Mini Unit: Claude Monet; Creation Science 2; Teens: Making My World a Better Place

#27 Feb 06

Vol 7 #27: Feb 06: Overseeing a Busy Mom's Life; Catching up & Cruise; Wright Photos; Soup Makes the Meal!; Proverbs 31: Spoil; SoundBites from Heaven; Self Esteem & Real Love; Adoption Promises; Weekly Wakeup: Home Atmosphere; Art: Teaching with Authority; Unit Yak: Birds; Mini Unit: van Gogh; Gift of Goals; Library Organization; Teens: Finding Your Voice

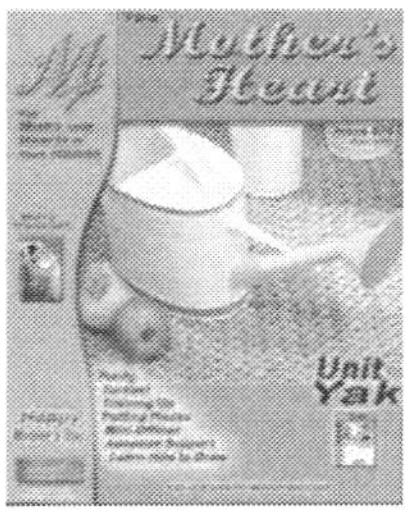

#28 April 06

Vol 7 #28: April 06: Potting Places & Lesson Plans; Unit Yak; Online Forum; Weekly Wakeup with Kym Wright: Training Up; Women: Living Life on Purpose; Bathroom Brush-Up; Easter Traditions; Purity; Pillars of Influence; Proverbs 31: She Does Him Good; Adoption Support; Special Ingredient: You; Art:

Learning How to Draw; Unit Yak: Turtle 101; Mini-Offices; Mini Unit: Mary Cassatt; Teens: My Top Five

#29 June 06

Vol 7 #29: June 06: The Bread of Life; Bread Recipes; Unit Yak; Online Forum; Weekly Wakekup with Kym Wright: Comparisons; Healthy, Fun Bread Recipes; Mommy, I'm Hungry!; Let's Talk Sex!; And The Garden She Grows; Proverbs 31: Working Delighted Hands; Adoption: A Heart Like His; Before You Adopt; Art: Evaluating Student Artwork; Unit Yak: Patriotism in Art; Mini Unit: Renoir; A Framework for Family Fun; Keeping Track; Teens: Stereotypes

#30 Aug 06

Vol 7 #30: Aug 06: Weekly Wakeup: Do It!; When Can I Date? Proverbs 31: Merchant Ships; Adoption: Legal-Ease; Studying Art History; Unit Yak: Vote for Credit; Zoo Adventure!; Mini Unit: Carl Larsson; A Creation Science Camp; Teens: Stars in My Eyes

#31 Oct 06

Vol 7 #31: Oct 06: Mediterranean Refreshment & Cuisine; e-Couragement from Kym: Unik Yak, Weekly Wakeup, and The Mother's Journey; Women: Living Life on Purpose CD Series & Books: Children & Chores, Family Meals, Moms Mentoring Moms; Haven for the Hurting; God's Provision; Groceries the Wright Way; Sweet Smells of Autumn; Bridging the Generation Gap; Keeping First Things First; Rising Early - Another Side; Saying Goodbye to Your Wife; Studying Art History; Unit Yak: Making Units Yours: Just a Leaf; Mini Unit: Michelangelo; College: Get a Jumpstart on Life! Pageant Prep: Hair, Makeup & Tech

#32 Dec 06

Vol 7 #32: Dec 06: The Merriest of Christmases; Wright Photos; e-Couragement from Kym: Volunteer Unit; From around the World; Kym's Talks: Moms Mentoring Moms, Family Meals, and Children & Chores; Mommy as Healthcare Giver; Finishing Strong!; A New Beginning; The Garden; A Hot Cup of Humility; Infertility: Make the Exchange; Art Journals; A Homeschool Chrismon Christmas; Obtaining Services for Special Children; Unit Yak: Making Units Yours: Introduction to Microscopes; Microscope Adventure!; Prom Prep

#33 Feb 07

Vol 8 #33: Feb 07: Burn-Out; Healthy Snacks; Preparing for Baby #8; God's Provision; Serving Our Families, Morgan's Bread Cookbook; Why Schedule? Changing Our World; Teaching Art: What to Expect of Different Age Levels; Family Garden Dreams; Garden Books & Tools, Catalogs, Month-by-Month Checklist; Homeschooling in the Maze of Autism; Mini Unit: Marc Chagall; Pageant Prep: The Box; February First Aid.

#34 April 07

Vol 8 #34: April 07: Whole Wheat Bread Machine Mix and Recipes; Healthy Cinnamon Buns; Simple Survival Schedule; Dana's Wise Mother; What Pastors' Wives Wish Everyone Knew; Keeping the Love in Discipline; Imaginations in My Mind; Art: Studying from Nature; Mini Unit: Paul Klee; Picturing a New Code for Spelling; Teens: My Experience in Romania; Prom Prep.

#34 April 07

Vol 8 #34: April 07: Commitment to Sisterhood; Art: Visiting a Museum; Crossfire; Library Course for Credit; Special Needs Children: Train Them Up; The Best (School) Year Ever!; The Schizophrenic Faculty; Occupying Preschoolers; What Is Giftedness?; A *Good* Government Program?; Proverbs 31: Big Cook; Pastor's Wife: Balance Life; Pageant Prep: Technical; Punching the Clock: Time Management for Moms?; Time: An Irreplaceable Commodity; Milled Flour from Whole Grains

Hands-on Unit Studies from
Learn and Do

Microscope Adventure!

With hints on buying a microscope to how to use it effectively, this unit covers it all! The $1 Rule of Optics, people to study, how to make permanent, semipermanent, well, wet and dry mounts. With the lab sheets included, use inexpensive items around the house, to study plant & animal cells, microbes, fibers, insects, crystals and more. Draw observations and fill in scientific data on lab sheets. 4th grade & up.

Photography Unit

Learn camera parts & types, along with up-to-date information on digital cameras. Understand lenses & film, composing pictures, focal point, lighting & angles. Practice photographing people, silhouettes, groups, portraits & children. Landscape, panorama, action shots, photographing animals, special effects & more. 4th grade & up.

Botany Unit

One of the best sellers in our catalog of studies. Learn all about plants from the roots up. Life cycles, needs, usage, differences, uniqueness, and comparisons. Microscopic to hands-on labs with plenty of Lab Sheets and flashcards provided. Appropriate for 6th grade through high school ~ this 160+ page study provides an exciting and comprehensive look at the plant world around us.

COLOR! Unit

with an interactive CD!: a terrific hands-on experience for learning color theory and its applications in fine art. This very interesting & entertaining unit will expose your students to color through exploration of fine art reproductions, the theory behind why certain color combinations work together and others don't. Full-color examples & hands-on labs, sure to capture the imagination of even the most kinesthetic student.

Library Unit

How to use the library, obtain a card, know the rules, learn etiquette and how the library is organized. Create personal reading lists, library notebook & library bag - all to save time at the library & improve our skills. Go online to look up books, reserve resources, and access interlibrary loans. Build a library, start a story hour, memorize the Dewey Decimal System, & earn high school credit for the course. 3rd & up

Bird Unit Study

Dubbed "Unit Studies for the Clueless" by Mary Pride's reviewer, it includes everything needed for a comprehensive study of birds. With lots of hands-on activities, it's a "no-brainer" - for mom/teacher, anyway. Sit back and learn along with the students. Identify birds in the field, air, or on water. Research projects help you learn bird species, beaks, feet, habitat, and incubating eggs.

Turtles! Unit

with a phenomenal CD of an actual dissection!: We all love turtles - watching & feeding them. Dig deeper and learn turtle anatomy along with the difference between reptiles and amphibians. Can you tell a turtle's age by its shell? How long do they live? Dissection included, to see the fascinating way turtles are created. Learn how to open the shell and all the external and internal parts. Since there was so much information for the required FunSheets™, we split it into two age levels: Younger (Pre-K-4th) and Older (4th-adult). While they both include some pages the same (Turtle Report, Plant or Animal, Anatomy, Sea Turtles), the presentation is different.

Younger Turtle FunSheets™: a must-have to go with the Turtles! Unit, for ages pre-K-4th. Includes graphs, mapwork, 3 turtle diet booklets, turtle identification with stickers, turtle vs tortoise worksheets, life cycles, and sea turtles graphs.

Older Turtle FunSheets™: includes observation graphs, mapwork, chordates research, measuring turtles, complete dissection, turtle vs tortoise research, various names for turtles, taxonomical classification, dietary preferences charts and sea turtles research.

Hands-on Unit Studies from Learn and Do

Spiders Unit

90-pages & gives a solid foundation in spiders, systematically studying: taxonomy, internal and external anatomy, diet, life cycle, habits, habitats, venomous types, handling, observations, experiments, webs, silk, and more.

Victorian Sewing and Quilting

Learn to quilt, bead, make sachets and Victorian pillows. Study color, harmonies, and the color wheel's application to fabric. Sew with modern and antique lace. Especially appropriate for the aspiring homemaker ~ appropriate for 6th grade through adult.

Flower Arranging and Wreaths

Study the wonderful art of floral arrangement and crafts, along with making wreaths, all in a systematic approach. Colleges teach this type course as "Horticulture 352: Flower Arranging." Enough for a one-semester high school credit. With marketing plans & practice, it becomes a one-year high school credit.

Goat Unit

70+ pages of fun, learning and hands-on activities. Practice milking at home without a goat! Anatomy, genetics and practical raisin' requirements are discussed, researched and practiced. This study is a guide to assist you in getting to know goat breeds and needs. A marketing and business plan is also included. All ages.

Sheep Unit

covers literature, history, hands-on fun, practical how-to's, raisin' requirements, anatomy, shearing, washing fleece and craft ideas. Learn the breeds, needs and feeds of sheep. Learning and fun for the whole family. All ages.

Poultry Unit

covers chickens, ducks, turkeys, and geese. Learn about incubating, raising and breeding. Supermarket Dissection, Lab Sheets, and plenty of Fun SheetsTM are included. All ages.

Arachnid Unit

covers the study of scorpions, pseudo-scorpions, whip scorpions, mini whip scorpions, mites, ticks, harvestmen and others. Makes a great companion study to the Spiders Unit. Study arachnids, their taxonomy, size, habits and habitats. Activities, labs, questions and information help you discover all about harvestmen, mites, chiggers, and ticks as well! So, come along on an Arachnid Adventure!

Volunteer Unit

As our homeschool children turn into preteens and teens, we see a need to help them learn about the world in preparation for their upcoming lives as adults: hands-on experiences, to see the poverty, the need, and take part in bringing hope and healing. Working with established groups, creating your own, or volunteering individually, there are always opportunities for service. 4th grade & up.

Free e-Couragement™
from Kym

Get your weekly dose of free encouragement from Kym Wright!

An invitation from Kym:

One of my favorite verses about motherhood is when Deborah, a judge of Israel, and a woman of might in battle, sings a song to the Lord,

Until I, Deborah, arose, until I arose, a mother in Israel. Judges 5:7 b, c

That's my prayer for us, that we arise to become the moms God has intended for us to be. It won't look the same in each family, but it will be based on a similar foundation of love, respect, time, training, discipline, discipling, fun, encouragement and all of those other aspects we admire and need.

Join us?

From one mother to another,

Kym Wright

Welcome to the **Unit Yak!** a **free monthly e-zine** for unit study users – and those who are interested.

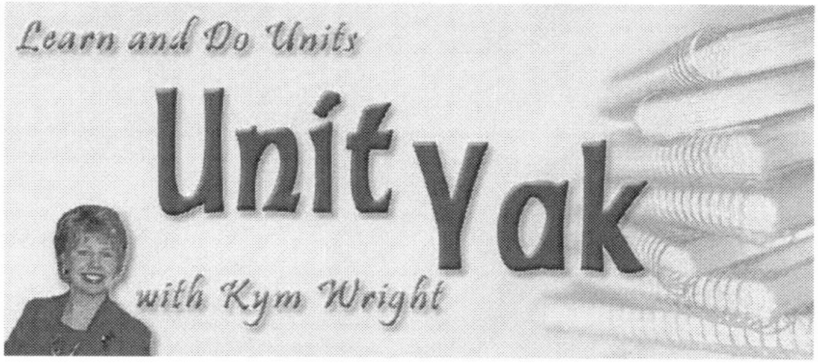

E-zine, E-Couragement, E-xactly!

The basic concept of unit studies is using one topic or literary selection to incorporate the majority of school subjects, especially: science, history, social studies and geography, writing, art and reading. Most times, math and grammar concepts are studied separately from the unit, but can be practiced within its context. There are primarily three ways to obtain unit studies:

- use what is available on the market
- heavily customize one that you find
- write your own

Kym shares a practical approach to schooling using unit studies, along with tips on how to make it work for you, your students, and your homeschool. Join Kym as she gives advice, and shares hints which make unit studies come alive. From the planning, gathering materials and supplies, record-keeping and actually teaching - Kym helps and encourages you along the way.

Learn and Do Unit Studies

Qty	Description	Print	eBook	Info	Total
	Arachnids Unit Study	$17.95	$15.95	Info	$
	Bird Unit Study	$17.95	$15.95	Info	$
	Botany Unit Study	$21.95	$19.95	Info	$
	Extra Botany Flash Cards	$8.00	N/A	Info	$
	Extra Botany Lab Sheets	$8.00	N/A	Info	$
	Bread Baking Unit Study	N/A	$9.95	Info	
	Color Unit Study w/CD	$17.95	$15.95	Info	$
	Flower Arranging & Wreaths	$14.95	$12.95	Info	$
	Goat Unit Study	$13.95	$11.95	Info	$
	Library Unit Study	$19.95	$17.95	Info	$
	Microscope Adventure!	$17.95	$15.95	Info	$
	Extra Microscope Lab Sheets	$8.00	N/A	Info	$
	Photography Unit Study	$15.95	$13.95	Info	$
	Poultry Unit Study	$15.95	$13.95	Info	$
	Sheep Unit Study	$13.95	$11.95	Info	$
	Spider Unit Study	$17.95	$15.95	Info	$
	Turtles Unit Study w/CD	$15.95	$23.95	Info	$
	Turtles Activity Pack - Older	$8.00	included	Info	$
	Turtles Activity Pack - Younger	$8.00	included	Info	$
	Victorian Sewing	$15.95	$13.95	Info	$
	Volunteer Unit Study	$19.95	$17.95	Info	$
	Organizing the Queen (Pageant help)	$7.95	$7.95	Info	$
				Unit Studies Total	$

Living Life on Purpose

Qty	Women: Living Life on Purpose	Print	eBook	Info	Total
	Living Life on Purpose 3rd Edition	$19.95	N/A	Info	$

CD Sets & Books

Qty	Description	Print	eBook	Info	Total
	Magazine Article Writing 6 CD Set	$39.95	N/A	Info	$
	Family Meals 3 CD & Book Set	$19.95	N/A	Info	$
	Children & Chores 3 CD & Book Set	$19.95	N/A	Info	$
	Creating Creativity CD	$8.95	N/A	Info	$
	Moms Mentoring Moms CD	$8.95	N/A	Info	$
	Gift of Goals CD	$8.95	N/A	Info	$
				Living Life on Purpose Total:	$

The Mother's Heart Magazine

(formerly *Open Arms Magazine*)

Qty	Description	Print	eBook	Info	Total
	1 Year e-Subscription (6 Issues)	N/A	$21.95	Info	$
	2 Year e-Subscription (12 Issues)	N/A	$39.95	Info	$
	Back Issues (eBook)	N/A	$4.00	Info	$
	Yearly Anthology (eBook) Check one	N/A	$21.95	Info	$
	❑#1 ❑#2 ❑#3 ❑#4 ❑#5 ❑#6 ❑#7			**The Mother's Heart Total:**	$

Make Check Payable to:
alWright Publishing
PO Box 81124
Conyers, GA 30013
Postage included in price.
US Funds, please.

Unit Studies Total:	$
TMH Magazine Total:	$
Living Life on Purpose Total:	$
Total Order:	$